LYD

A New Lynton & Barnstaple Locomotive

The Story of the Re-creation of a Lost Class of British Narrow Gauge Locomotives

Lyd enters Minffordd on the Ffestiniog Railway with an Up passenger train
(from the painting by the late David Perrin)

LYD

A New Lynton & Barnstaple Locomotive

© *James Evans, Paul Lewin,*
David Payling and Jon Whalley

First Edition, May 2011
ISBN 978 0 901848 09 3

Published by The Ffestiniog Railway
Harbour Station
Porthmadog LL49 9NF

Designed and Printed by
Gwasg Eryri - Snowdonia Press
Unit 3, Ystad Ddiwydiannol Penamser
Porthmadog LL49 9NZ

2011

Front Cover: *Lyd, in British Railways mixed traffic black livery, leaves Cutting Mawr with an Up train for Porthmadog on 20th April 2011.* (Roger Dimmick)
Contents page: *A hybrid taken from a photograph of Exe and the SR General Arrangement drawing of Lew* (Peter Dennis, Lyd Fund)

Contents

- I -
Foreword

It was in 1964 that I started my personal Lynton & Barnstaple Railway Society, which comprised my brother and a couple of school friends. Having absorbed every detail in LT Catchpole's evocative history of the line, we set off to explore what remained. We walked most of the track-bed, still quite passable except where river bridges had been demolished, and were only thrown out of one man's garden. We came across a complete set of perfect carriage doors, collected bits of locomotive coal and telegraph insulators, and took photographs with a Box Brownie camera. We made friends along the way and were interviewed by the local press. We were even invited to a showing of some amateur movie film of a train in action. I was hooked! What a charming transport backwater, set in the most beautiful, romantic, rolling North Devon countryside! It had been a pioneering, environmentally sensitive, public facility, though the station at Lynton was undoubtedly too well hidden from view. Grandiose it was, with enough land for double track, full signalling and modest curvature considering the terrain. Even the goods stock had been vacuum braked and the locomotives had been a charming and beautifully presented class of 2-6-2 tank engines built by Manning, Wardle & Co., of Leeds. They were the vanguard of a type subsequently developed by the GWR for branch line work all over the West Country. In its early years it had even pioneered a feeder bus service but, although that was killed by legislation, the vehicles were later used by the GWR for the first successful integration of road and rail services. Advice had been sought from the Ffestiniog Railway, leading light of narrow gauge lines, but whilst the FR gauge was adopted, track specification was reduced and the genius of Fairlie's Patent was not appreciated.

The end for the Lynton & Barnstaple (L&BR) came after just 37 years of service, and it was the first time a complete, well maintained, part of the national network had been closed outright and totally destroyed within months. The railway world was shaken by this premonition of Beeching. The feeble-minded still maintain that rural lines could never have been made to 'pay' or 'compete' with road transport, but when authorities like Devon County Council spent fortunes building bypasses such as that around Parracombe, the playing field was tilted in favour of the subsidised bus and motor car. Surely the railways could have had a role in an integrated system. As one thinker pointed out: "lamp posts are jolly useful even though they don't make a profit!" When the line opened, road vehicles spluttered along at eight miles per hour in the dust and the mud (my grandfather was fined for 'speeding' at little more!) and twenty miles per hour to Lynton must have been a wonder.

Thoughts of building a replica Manning, Wardle were born out of frustration over the slow progress with restoring my two favourite railways, the L&BR and the Welsh Highland Railway (WHR), but lack of drawings was an obvious problem. After much correspondence, I located an original set of 1897 drawings, but they were in poor condition and

being laboriously redrawn by a model maker. The break came when I discovered that a friend's father had worked at Eastleigh and had purchased the SR tracings for Lew, which were about to be burned along with about ten tons of steam era drawings in the works yard. Ten years later a complex deal was agreed and I had the drawings, but the estimated construction cost had more than doubled. My last excuse for delay was the outside possibility that Lew, the 'Lord Lucan' of the narrow gauge, might suddenly reappear. I had always taken an interest in attempts to discover her fate and an announcement in September 1991 that she had been FOUND came three days before my brother's engineering works was to have started on the replica; after fifty years! Subsequent enquiries convinced me that earlier reports that she had been scrapped about 1957-8 were probably correct.

The Ffestiniog Railway Hunslet Hundred Gala in 1993 was a turning point for Lyd as, having had a wonderful weekend running Velinheli (the prototype Alice class quarry locomotive) and having found a great deal of enthusiasm for Lyd, I rejoined the FR Society after a long break. After teaming up with Paul Lewin to help finance Linda's boiler overhaul, Paul negotiated acceptance of my efforts thus far, and Lyd became a Ffestiniog Railway project.

One question that I'm often asked is: "Why is the locomotive being built in North Wales and not in North Devon?" Indeed the Ffestiniog was even accused of stealing the Lynton & Barnstaple's project! There have also been criticisms for not starting the project years earlier and for something called 'mission creep'.

All of these points can be answered by saying that projects of this kind only happen when the right people come together at the right time. Whilst I was pottering away quite happily in my own workshop, at a pace I could just about finance, it was entirely down to the Ffestiniog Railway, and the number of people who came forward with commitments of cash and practical assistance, that Lyd accelerated into public reality. As for the mission, this has remained almost totally unchanged since my original tentative suggestions in the Lynton & Barnstaple Magazine.

At the outset, the project team anticipated that Lyd would be tested and commissioned as a complete locomotive in 2001. However, that was not to be. Difficulties over the availability of workshop space, engineering skills, and funds combined to push the time scales back. Now, at last, with the locomotive complete we are beginning to explore the potential of this classic design.

Finally, I thank all who purchase a copy of this book. The proceeds will go entirely to supporting Lyd

James Evans

- II -
Introduction

The three locomotives built by Manning, Wardle & Co. in 1897 for the opening of the Lynton & Barnstaple Railway (L&BR) can lay claim to be among Britain's best loved narrow gauge locomotives. With Lyd now complete it represents a unique class of narrow gauge engines which had become extinct (with the possible exception of Lew), having been the subject of a project to recreate one for service in the 21st century.

Who could see the photographs of one of the engines with its train traversing the Exmoor foothills and remain unmoved? Or the photo by FE Box of Taw and its train leaving Barnstaple and passing Rolle Quay? Not to mention the idyllic scene by LT Catchpole, first historian of the railway, of Lew taking water at Parracombe, as the guard walks forward for a word with the driver?

The sheer photogenic beauty of the L&B and

Lew leaves Lynton with a Barnstaple-bound early evening train in the Summer of 1935. (GH Soole, NRM)

the fact that, as a popular, well-known and singular holiday route, it was indeed much photographed, has resulted in it being one of the most loved of all railways amongst model makers. These include makers of locomotives in all gauges and also those who make complete layouts with buildings and landscapes in great detail.

Another reason for the abiding, and probably growing, fascination with all things L&B, even some 79 years after the end of its life, is that it died young, tragically and still beautiful. As with the loss of some especially gifted or attractive individual, many are left wondering 'what if' they had survived a little longer? In the case of the L&B it has often been suggested that, had the line been spared for another couple of years, perhaps only as a summer service for the benefit of holiday makers (the winter service used by ever fewer locals was undoubtedly loss-making), then with war looming the line might well have been retained through to 1945. There

LEW *takes water at Parracombe on 8th August 1935.* (LT Catchpole, Oakwood Press)

LEW *passes a Barnstaple bound train at Chelfham as it heads for Lynton with a Down train in 1935.* (NGRS Archive)

is then just the possibility that, with the return of peace, but the somewhat slower return of prosperity, it might have scraped through to join the rebirth of the Talyllyn and Ffestiniog Railways in the 1950s.

Had this occurred there is little doubt that the Lynton & Barnstaple Railway would be today one of the premier tourist attractions of the West Country, if not of the United Kingdom.

An elevated view of *Lew taking water at Chelfham in 1935.* (AB MacLeod, JDCA Prideaux Collection)

Lew stands outside Pilton running shed in Barnstaple awaiting its next duty in Summer 1935. (NGRS Archive)

Lew and train are near Snapper Halt in this 1935 view. (JDCA Prideaux Collection)

With the cowcatchers perched on the boiler top and chalked with the auctioneer's lot number, Lew stands at Barnstaple Town in early 1936, about to leave Devon to begin its new life overseas. (NGRS archive)

- III -
The Manning, Wardle Tank Engines

~ *The Time Perspective* ~

The Manning, Wardle tank engines for the Lynton & Barnstaple Railway were highly distinctive in their appearance. They showed little similarity to locomotives already at work on the British narrow gauge. But, what was the background to their design, and is it possible to place them in the perspective of development of designs for the narrow gauge? Looking back to 19th century narrow gauge practice in Britain, a common arrangement was for locomotives to employ a link motion located between the frames. This was usually coupled with outside cylinders in a horizontal layout. The benefit of this was to give a short steam circuit from boiler to valve chests and from cylinders to the blast pipe. The Ffestiniog Railway locomotives, and the slightly later Hunslet quarry locomotives of this period, are examples. The same was true of the first locomotives of the North Wales Narrow Gauge Railways (NWNGR). On the standard gauge it was common practice for locomotives to have both inside valves and cylinders, a compact arrangement and one which retained heat better than outside cylinders. On the narrower gauges, however, there was generally insufficient space between the wheels to accommodate both cranks and eccentrics on a single axle.

An early sign of changes to come was the Hunslet design for the Penrhyn Quarry Railway (PQR) main line locomotives - Charles of 1882, and Linda and Blanche of 1893. Here the movement of the link motion, mounted in the traditional way, was transferred via rocking levers and shafts from between the frames in order to control valves that were mounted outside the frames. In this case the steam chest and cylinder shared single castings, one on each side of the smokebox. A similar layout, with outside valves placed above the cylinders, was employed by some of the British main-line railways at the turn of the 20th century. There were, for example, the pioneer Churchward designs, including his Saint class and other two-cylinder locomotives. For them the movement of the inside Stephenson link motion was transferred outside the frames via rocker shafts to control the valves. The need for a pit between the rails to gain access to lubricate the motion was offset to some extent by the fact that the coupling and connecting rods were very easy to remove, if necessary, through the absence of outside valve gear.

The PQR and GWR arrangements marked something of a half way house in moving away from the inside valve gear and outside cylinder arrangements common in 19th century locomotives. A new pattern however began to emerge at the beginning of the 20th century. This was driven by the need for working economies, including the ability to service locomotives from the trackside rather than requiring a pit between the rails. The result was for valve gear, valves and cylinders all to be placed outside the frames, making the whole more accessible for oiling round and for routine maintenance.

Thus, for the 19th century narrow gauge, the large 0-6-4T engine Beddgelert had been built by

Hunslet in 1878 for the NWNGR. It was equipped with inside valves and motion to control outside cylinders. Thirty years later the 2-6-2T Russell was built to similar dimensions and power output by Hunslet to run on the same railway. Now, however, the locomotive was equipped with outside cylinders and valves which were controlled by outside motion. Russell and the later Hunslet single Fairlie Gowrie, not to mention the Welsh Highland Railway's (WHR) Baldwin, No.590 and the FR's Alco, Mountaineer, brought external Walschaert's valve gear to the fore, controlling outside valves and cylinders. The new WHR's South African NG15s, NG/G16s and the Tasmanian Garratt, K1, too, are also fitted with this motion. This perhaps highlights one of the more unusual features of the L&BR locomotives, their external Joy valve gear. Few narrow gauge locomotives on the Welsh narrow gauge were so equipped (the Moel y Gest quarry's vertical boiler 0-4-0VBT being an interesting exception, possibly following marine practice).

The 1897 Manning Wardle works photograph of Exe *shows the locomotive in its original condition with full motion covers and extended cab side sheets in front of the spectacle plate.* (RL Knight Ltd, James Evans Collection)

Manning Wardle was within months of final closure when Lew *was completed in 1925. Unusually the works photograph shows the engine in the customer's livery rather than photographic grey.* (NRM)

Essentially the basic design of the L&BR locomotives, with outside cylinders and outside frames, would have offered the same ease of fitting of Walschaert's valve gear as of the Joy radial motion, although a link motion such as Stephenson's, with its two eccentrics per valve, would have been more complex. The large tank locomotives built only a few years later for the Portmadoc, Beddgelert and South Snowdon Railway (PB&SSR) (Russell, 1906), the Welshpool and Llanfair Light Railway (W&LLR) (1902) and the Leek and Manifold Valley Light Railway (L&MVLR) (1904) all had outside frames and cylinders like the L&BR locomotives of 1897. They all carried outside cylinders, valves and Walschaert's valve gear from the start. The two Vale of Rheidol Railway (VoRR) engines built by Davies & Metcalfe (1902) were an exception, having a most unusual and impractical valve gear, but were later replaced by three Swindon-built locomotives to the more modern pattern. The L&BR locomotives can thus be seen, within this perspective, as one of the earliest designs to employ what was to become a standard layout for narrow gauge steam designs for the 20th century.

It should be pointed out in passing and also to explain the adherence of the GWR to what might appear to be a more antiquated layout, that Walschaert's, Joy's and other radial valve gears provide fixed lead, i.e, valve opening to steam when the crank is near front or back dead centres. This feature tends to determine whether a locomotive is a slow-speed 'slogger' or a high-speed racer! However, with the link motions the lead varies at different settings of cut-off, the shorter cut-offs generally being used at higher speeds. Churchward, by a very clever bit of geometry, managed to arrange for no, or even negative, lead in full gear and increasing lead at shorter cut-offs. Thus his locomotives were capable of both 'cart horse' work with heavy coal trains and also free running at speed with remarkably short cut-offs.

Few other Manning, Wardle engines resembled Yeo, Exe and Taw (1897). The two Cape Government Railway locomotives of 1902 were an exception with their very similar side elevation. A boiler mounted sand box was also used, but Walschaert's valve gear replaced the Joy motion of the L&BR engines. (FWH Collection)

~ *Design Influences on the L&BR Locomotives* ~

The appearance of the Lynton & Barnstaple Railway locomotives of 1897 was quite distinct from other earlier UK designs for the narrow gauge. They also showed little family resemblance to earlier engines constructed by Manning, Wardle! What were the influences on their design? They might include the specification supplied by the railway's consulting engineer to potential builders. They might also include the influence of leaders in narrow gauge engineering of the time, similar to the input of Robert Fairlie and Charles Spooner some thirty years earlier. Thirdly the other locomotive engineering companies in Leeds might have had their own role, directly or indirectly, in what appeared.

A recently published survey of Manning, Wardle's output shows that no locomotives built earlier than 1897 had been similar in general layout to the L&BR tank engines. Nor have any so far come to light from this period as being fitted with Joy valve gear. In later years, however, Manning, Wardle did build a few locomotives which were of similar dimensions and appearance. These especially included the two 2-6-4T locomotives built for the construction of the Avontuur Railway in the Cape region of South Africa in 1902. In contrast to the L&BR engines the large, sturdy tank engines for the Avontuur line were fitted with Walschaert's valve gear and connecting rods driving onto the trailing coupled axle rather than the centre axle. They were not provided with motion covers, as used for the L&BR, but were equipped with cowcatchers, and also with one of the unusual rectangular cast boiler top sand boxes, as fitted to the L&BR engines.

~ *The L&BR Consulting Engineers* ~

On the Lynton & Barnstaple Railway in 1896 and also on the Vale of Rheidol in 1902 the consulting engineer was Sir James Szlumper. Sir James had been trained as a civil engineer, his first appointment being as the assistant to the civil engineer for the Manchester & Milford Railway. After the termination of this project he served as the civil engineer to the construction of other railways, especially in Wales - the Barry Railway and the Aberystwyth to Carmarthen railway being two. He was thus professionally unlikely to have strong views on the mechanical engineering issues of locomotive design. In fact it was William Szlumper, his half brother, who was responsible for the design of rolling stock on both the Lynton & Barnstaple (1897) and the Vale of Rheidol (1902) lines.

The research of Mr CC Green, historian of the Vale of Rheidol, has produced some very interesting information on VoRR locomotive design. According to Mr Green the Invitation to Tender to build the VoRR engines was accompanied by a performance specification and an outline drawing of the locomotive by William Szlumper. The rough drawing showed a locomotive of very similar appearance to the L&BR design of 1897. The provision of a rear bunker was omitted, however, being replaced by bunkers on either side of the firebox. Interestingly, too, there were no cowcatchers. Unfortunately the 1896 invitation to tender for the L&BR locomotives has not so far come to light. The locomotives which eventually appeared did, however, represent a departure from the general appearance of Manning,

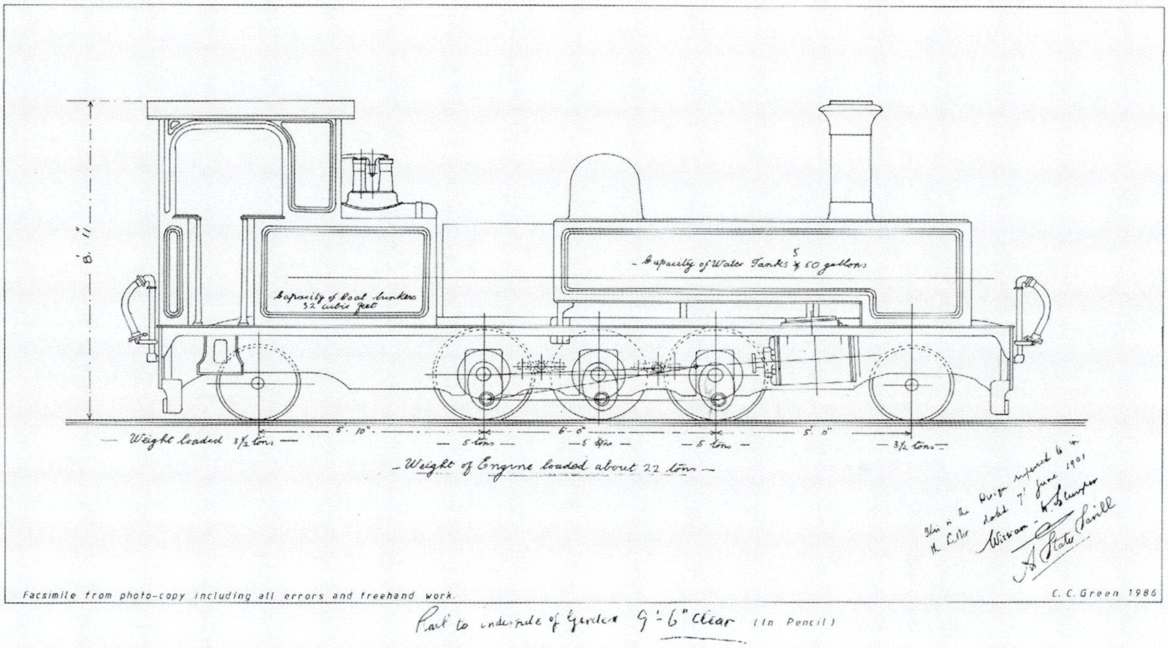

W.W. Szlumper's tender drawing of a Lynton & Barnstaple style engine. The draughtsmanship has all the marks of the haste in which it was most obviously prepared.

The tender documents issued in 1902 by W Szlumper for locomotives for the new Vale of Rheidol Railway included this drawing. It shows a strong family resemblance to the L&BR locomotives of 1897, although with a better length for the connecting rods. There was no back bunker and the tanks were moved forward, both features making for an improved weight distribution. (Wild Swan Press)

Two locomotives were built for the VoRR by Davies & Metcalfe, appearing in 1902. Their link motion was concealed between the outside frames and the wheels, an inaccessible arrangement. (Great Western Trust)

Wardle locomotives of that time. The question as to whether the Manning, Wardle locomotives were specially designed for this railway, or represented a modification to a new and speculative design, or even part-built locomotives, must therefore remain conjecture.

Whilst the Manning, Wardle engines for the L&BR had been built with external Joy valve gear, the instructions to potential bidders for the VoRR engines expressly excluded the use of outside valve gear! The reason for this restriction is not clear but it was resolved by the successful bidder, Davies

Left: *The GWR built three locomotives in 1923-4, similar in general dimensions to the earlier Davies & Metcalfe engines for the VoRR. The exception was the motion, Walschaert's valve gear being used. Nos. 7 and 8 stand in the running shed at Aberystwyth in 1986.* (Hugh Ballantyne)

Bellow: *Vale of Rheidol No.8 receives an overhaul in Swindon works in May 1954. The partly dismantled state of the engine allows a clear view of its Walschaert's valve gear.* (Hugh Ballantyne)

& Metcalfe, by fitting an ingenious arrangement of Gooch link motion. It was sandwiched into the very tight and inaccessible space between the outside frames and the wheels, and was connected via rocking shafts to the outside valves. The valve chests were cast integrally with the cylinders.

The designs for the L&BR locomotives tendered by Hunslet have survived. The research of Don Townsley and Geoffrey Horsman in the Hunslet archives in Leeds has revealed that the Hunslet options included 4-4-0 and 2-4-2 tank engines, all with outside frames, cylinders, and valves but with inside valve gear operating the outside valves via rocking shafts. This was a similar design solution to that offered by Baldwins of Philadelphia for Lyn (1898), to meet the performance specification set by the Lynton & Barnstaple. The Hunslet engines were drawn showing cowcatchers as an option. The profile of the locomotives is typical of Hunslet designs of the period and shows no similarity in appearance to the eventual Manning, Wardle design. Rather contrarily, taken together, this perhaps suggests that inside valve gear was to be used, as in the case of the VoRR.

Later, in 1923-1924, the GWR built three new locomotives for the VoRR. They were similar in appearance and specification to the 1902 Davies & Metcalfe design. However the new engines were fitted, not with the inaccessible and vulnerable original valve gear arrangements, but with a very neat version of outside Walschaert's valve gear. It has been said to be closely related to that fitted to the Swindon steam rail motors of the time.

From the above it is clear that William Szlumper took an active part in the design, at least in outline, of the L&BR and the VoRR locomotives. The reason why such different arrangements for valve gear were eventually used however is less clear. Certainly the L&BR engines were not criticised on any grounds that could be the result of their use of Joy valve gear.

(James has concluded that the very long full-gear cut-off may have been the reason for selecting Joy's valve gear: useful in re-starting a train on 1 : 50 gradients, and probably the reason why it is also used on mountain rack locomotives.)

~ *ER Calthrop and the Barsi Light Railway* ~

A perhaps more likely source of influence on the design of the Lynton & Barnstaple locomotives could have been Mr ER Calthrop. In the 1890s he had become an articulate and influential voice in narrow gauge and light railway policy. He had been trained as an apprentice engineer under FW Webb by the L&NWR at Crewe works, leaving in 1879, the year of publication of Joy's patent. He had probably not, therefore, been exposed to any work on Joy's valve gear during his apprenticeship. His next employment was in the Great Western Railway workshops at Swindon. In his long-term career as a consulting railway engineer Calthrop built up and maintained a productive relationship with two Leeds companies. They were the Leeds Forge, for rolling stock construction, and Kitson for locomotives. In 1897 he published a pamphlet on Light Railway Construction to outline his views. These views had been exemplified in the building of the Barsi Light Railway in India, of which Calthrop was the chief engineer. As such he was responsible for the specification of its equipment. The first Barsi Railway locomotives had been completed by Kitson in Leeds in 1896. The rolling stock had

The Newlay Exhibition of October 1896 allowed ER Calthrop to display and test the newly-constructed 0-8-4T Hamilton, *built to his specification by Kitson for the Barsi Light Railway in India.* (Birmingham City Museums)

In 1904 Kitson completed two 2-6-4T engines for the opening of the Leek and Manifold Valley Light Railway. The works photo shows them constructed with motion covers although it is believed that they these were removed before the locomotives entered service. (NRM)

been built in Leeds too, largely by the Leeds Forge. This company was a specialist in the application of the newly-developed pressed steel technology to building railway vehicle chassis and bodywork.

Calthrop, Kitson and the Leeds Forge took advantage of the completion of much of the rolling stock for the Barsi Railway to organise an exhibition in October 1896. This of course took place just as the moves to equip the proposed L&BR were gathering momentum. The intention was for the exhibition to serve as a shop window both for Calthrop's ideas and for the products of the two companies. The location was at Newlay, adjacent to the Midland Railway line between Leeds and Bradford. A selection of the new rolling stock for the Barsi Railway, together with the Kitson 0-8-4T locomotive Hamilton, were put through their paces on a specially constructed layout of two foot six inch gauge track. Calthrop's design concepts for the locomotives included a uniform and low axle load, to match that of the wagons and carriages under load as far as possible. They were to be equipped with outside frames with outside cylinders and valves, Walschaert's valve gear and long connecting rods. The up and down forces on the driving axleboxes of an engine with short connecting rods, such as the GWR large Prairie tanks, for example, made them prone to rolling when working on long cut-offs and at low speed. The Barsi locomotives were thus among the very first to display what were to become the long-term design preferences for larger narrow gauge steam engines in the 20th century. Taking place less than one year before the construction of the Lynton & Barnstaple engines, it seems more than possible that the ideas and concepts displayed at Newlay would appear in some measure in the L&BR engines. However, very little evidence has so far come to light on any connection or consultation between Calthrop and either the Lynton & Barnstaple board or Manning, Wardle & Co. Interestingly, Sir Thomas Wardle of Swainsley Hall, as a supporter and director of the Leek and Manifold narrow gauge line, would certainly have had contact with ER Calthrop during this period. Unfortunately, however, it is not known whether he had any connection with the eponymous Leeds locomotive builder.

~ *The Leeds Locomotive Builders* ~

Could the Manning, Wardle locomotive design for the L&BR owe something to the influence of the neighbouring Leeds locomotive builders such as Kitson, Hudswell Clarke, John Fowler or Hunslet? Certainly the Leeds engine builders are known to have worked closely together when appropriate during this period. The workshops of them all were located within about one square mile in the Hunslet district of Leeds. An initial contract for three locomotives for the L&BR placed with Brush of Loughborough had failed. As a result, the eventual order to Manning, Wardle, in late 1896, called for delivery in 1897, emphasising the need for haste. Although the engineering industry was in some disarray as a result of a major strike by the Amalgamated Society of Engineers, the delivery timetable for the three engines was honoured, the three engines being delivered some time before the railway was complete.

Taking first the case of Kitson, it is difficult to separate their designs from the specifications laid down by Calthrop, so close was their collaboration. These are embodied in the 0-8-4T engines of 1896 for the Barsi Railway. Some of the design features of these engines are reflected in the L&BR's 1897 Manning, Wardle 2-6-2T locomotives. They include the side profile, both designs showing long low side tanks and generously-sized workmanlike cabs with twin side windows, and a functional rear bunker with a flared top. The weight of the Barsi engines (29 tons, 8 cwt in working order) was spread over six axles, in line with Calthrop's specifications. Perhaps also reflecting this, the Manning, Wardle 2-6-2T engines spread their approximately 23½ tons over five axles (see notes to Appendix 1). The Baldwin Lyn of 1898 and the unsuccessful tender from Hunslet for locomotives of similar output to that of the 2-6-2T Manning, Wardle engines called for a four-axle (2-4-2T) layout. Further similarities included the use of motion covers and front and rear cowcatchers together with the patented Jones-Calthrop central buffer couplers. These couplings gave provision for locking and tensioning the chopper couplings. The Barsi engines differed, however, in that they were equipped with Walschaert's outside valve gear. Their sand boxes were neat fixtures on the leading ends of the side tanks (possibly vulnerable to damp weather conditions).

In Calthrop's 1897 paper he explained the reason for some of the features of the Barsi engines. A significant portion of this railway was constructed as a roadside line. The motion covers were to keep road dust out of the motion whilst the cowcatchers were necessary because of the large number of animals encountered on the unfenced railway. On the L&BR engines some of these design features enjoyed varying degrees of success.

First of all, the side-windowed cab, with its rear flared top coal bunker, included some less than practical features. The rear bunker proved to be unsuitable for carrying coal and is said to have been used mainly to house the fire irons. It has been suggested (James Evans) that the problem here was the already heavy load on the trailing pony truck axle (close to the limit for the journals and indicating that a four wheeled bogie may have originally been intended), combined with a lack of weight on the leading pony truck. A load of coal in the rear bunker, combined with the effect of the very stiff springing of the trucks, may have reduced the load on the front truck excessively. This in turn could have caused a proneness to derailment on uneven track, as was the case in the L&BR's early days, owing to very wet weather which had caused subsidence during the construction period. A similar problem was experienced with the PB&SSR Russell during its years as an industrial locomotive, eventually causing it to be operated with the leading pony truck removed! Perhaps for these reasons coal was carried instead in two small bunkers in front of the cab. Access to load them was hampered, however, by another feature of the cab design. The cab front (spectacle plate) was positioned, not at the leading end of the side sheets, but set further back at a point between the two side windows, making an awkward reach for filling the recessed coal bunkers. It also allowed steam to become trapped in the space in front of the cab, obscuring the driver's view. After a few years the cab was rebuilt to a more practical shape to avoid the problem. The rear bunker was enclosed within the cab at the same time. The newly improved access to space in front of the cab was used to enlarge the capacity of the front coal bunkers.

Turning to the motion covers and the cowcatchers, these had been included on the Barsi engines to make them suitable for roadside running. The L&BR however had been planned from the outset to be fully fenced! No section of the line involved street running, though Exmoor is deer and sheep country, the former being able to jump almost any fence and the latter able to escape down the line from any of the numerous accommodation crossings. Few other locomotives in Britain have been fitted with cowcatchers. In North Wales even the roadside Glyn Valley Tramway locomotives used skirts to enclose the motion, but had no cowcatchers. In Ireland too the practice was rare except on lines such as the Tralee & Dingle and the Cavan & Leitrim, where roadside running was a necessity. Calthrop's 2-6-4T locomotives for the L&MVLR, built by Kitson in 1904, omitted these features, perhaps confirming that he would not have specified them except for locomotives destined for a roadside railway. Interestingly the works photograph of the L&MVLR locomotives shows them to have been built with motion covers which were apparently then rapidly removed, possibly before the locomotives entered service. Thus, although the Manning, Wardle tanks were built with motion covers and cowcatchers, the former at least were of little practical use and indeed were soon pinned open and eventually largely removed. The protection fitted to Lew was vestigial

~ *John Fowler and Joy Valve Gear* ~

Two other very characteristic features of the Manning, Wardle tank engines were the Joy valve gear and the unusual rectangular boiler-top sand boxes. The location of the sand boxes was an advantage as the warmth of the boiler tended to keep the sand dry and free flowing. It has been pointed out that Joy's valve gear appears to have been employed by Manning, Wardle only for building the four 2-6-2T locomotives for the Lynton & Barnstaple. The sand box design was also only rarely used. In contrast, both of these details can be found in frequent use during this period at the adjacent workshops in Leeds of the John Fowler & Co. Steam Plough & Locomotive Works.

After the publication of David Joy's patent for his new valve gear in 1879 it was taken up swiftly by the L&NWR and the L&YR at their Crewe and Horwich workshops respectively. By the turn of the century there were some 3,000 locomotives in Britain running with this valve gear. Joy valve

A batch of 7 locomotives was built by John Fowler of Leeds for the Woolwich Arsenal Railway in 1885. The first five, including CORMORANT, *shown here, carried outside link motion.* (Museum of English Rural Life)

The final two Woolwich locomotives of 1885 were fitted with Joy valve gear and other features, such as boiler-top sand boxes, which were to appear later on the L&BR engines. (Museum of English Rural Life)

gear showed its greatest advantage when used with inside cylinders, the valves being in line vertically with the cylinders. The driving axle was left free of eccentrics, thus leaving the maximum length for the cranks and axle boxes. However, at Crewe in 1882, Francis Webb produced a variant for use outside the frames. In it the anchor link between the valve gear and the motion bracket of the original design was replaced by an eccentric rod and a short return crank mounted on the driving axle crank pin, as shown in the diagram. It was this variant that was taken up by John Fowler of Leeds and eventually used by Manning, Wardle on the L&BR locomotives.

The output of John Fowler & Co. Ltd during this period included a standardised range of railway locomotives. Their contemporary catalogues show at least eleven types for different gauges and power outputs. At least seven of these standard designs included the option of Joy valve gear, generally fitted outside the frames. It was always referred to in the catalogues as Joy's patent valve gear. Replacement die blocks to slide within the curved links (or slides) were offered at fifteen shillings per set in the 1893 Fowler catalogue. Michael Lane, the historian of John Fowler & Co., writing about this period, referred to 'David Joy's valve gear, most Fowler locomotives employing this locally inspired and patented system at this time.' This refers to the fact that the business of David Joy's family (W&E Joy & Co.) was also located within the same area of Leeds. Their speciality was the supply of steam lubricating oils. In 1885, John Fowler produced a batch of seven 0-4-2T engines for the eighteen-inch gauge Royal Arsenal Railway (RAR) at Woolwich. The last two of the batch (works Nos. 5063 and 5064) were named Owl and Pelican and were turned out, not with outside link motion like the first five, but with the return crank modification of external Joy valve

Fowler of Leeds used this variant of Joy valve gear on many locomotives, but Manning Wardle used it possibly only for the L&BR contract. Their names for the various levers are shown here and differ from those used by Fowler. (David Payling)

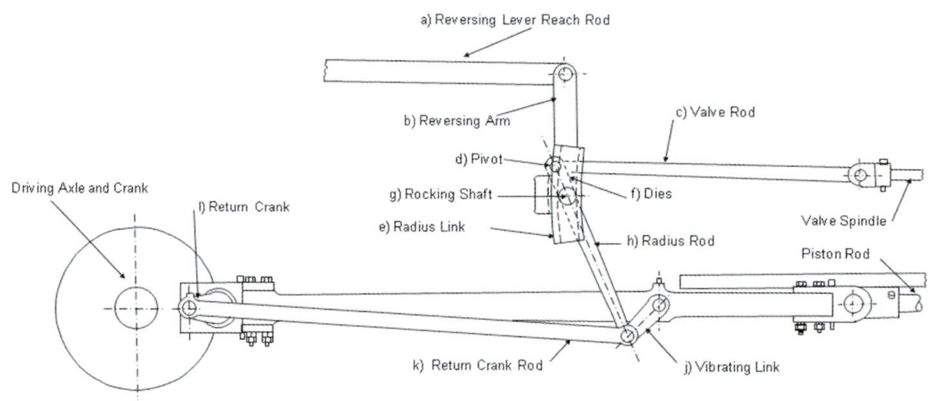

gear, essentially the same as on the L&BR tanks. This was only a year later than the building of the last Webb LNWR locomotives to be fitted with this version of Joy's motion, rather than an anchor link. The works photograph of No. 5064 shows another interesting similarity to the L&BR engines. They had been equipped with a pair of the same type of unusual rectangular sand boxes mounted on the boiler top, in front of, and to the rear of the dome. One of the disused Fowler locomotives rescued recently from a sugar plantation in Mozambique and brought to Britain, is fitted with external Joy valve gear (Fowler 0-6-2T No. 12817 of 1911, Marromeu Estate No. 4). Two others now awaiting restoration for the proposed museum on the Vale of Rheidol Railway at Aberystwyth are similarly fitted (0-6-2T 15515/1920 ex-Mozambique and 0-6-0T 10249/1905 ex-Fiji).

Thus one of the Leeds locomotive manufacturers was making extensive use of FW Webb's development of Joy valve gear within three years of its first use at Crewe. They also used the rectangular boiler top mounted sanding equipment as fitted to the 1897 L&BR engines. Such sanding gear appears to have been used by Manning, Wardle only on a few locomotives including the L&BR engines and the two 2-6-4Ts of 1902 for the Avontuur Railway, according to the photographic records in FW Harman's study of the company's output.

From the foregoing it can be seen that some of the features of the Manning, Wardle tank engines may have derived from the influence of the L&BR's consulting engineer, William Szlumper. It seems more than probable that ER Calthrop's views on light railway engine design were also a factor, certainly so in the case of the draw gear. The Barsi railway locomotives produced by Kitson exemplified these principles. They were exhibited at Newlay in 1896, a critical year in the evolution of the L&BR. Finally the use of the Webb variant of Joy's valve gear and the boiler mounted sanding arrangements were both standard features of the output of John Fowler at the time.

The valve events were checked using a full size metal model. (Paul Lewin)

- IV -
Enhancing The Breed

~ The Boiler ~

One of the limitations of the L&BR engines, as originally designed, was making insufficient steam. For this reason they were permitted no more than four carriages before double-heading was obligatory. This did, however, give some margin for picking up freight vehicles at intermediate stations. Considerable effort, therefore, has been made to improve steam raising.

The original Manning, Wardle boiler was constructed with 7/16 in. thick plate, a copper firebox with crown girders and a barrel in two rings, reducing in diameter towards the front, and all of riveted construction. Due to the small diameter of the front tube plate there were relatively few tubes (75) though they were of considerable length (120 in.). The firebox had an ample grate area of 8¾ sq. ft. and no brick arch was provided. The working pressure was 160psi of saturated steam, there being no superheating. The fuel was coal.

The design of the new boiler was influenced by the following factors:-

a) **Boiler plate of 7/16 in. thickness is no longer available.**

b) **The axle load on the trailing truck could be seen to be near the normal limit for journals (bearings) of the size provided, whilst that on the leading truck was so low that a ballast weight had been required (1,300lb, estimated).**

c) **Research indicated that the original locomotives were indifferent steamers and the limiting factor was the poor 'free gas area,' meaning the total cross-sectional area of the inside of the tubes, relative to the grate area.**

d) **Whilst a copper firebox had advantages when used with untreated acidic water, the disadvantage of its greater thermal expansion than the steel outer shell, difficulty of repair (especially with dry-lap seam problems), electrolytic activity round the tube ends and high first cost, mitigated against its use when water would be treated and oil firing had been specified from the start.**

In conclusion it was therefore decided to take advantage of using ½ in (12.5mm) steel plate throughout (except for tube plates (16mm) and to increase the working pressure to 200psi, thus enhancing thermal efficiency and boiler reserve. The general increase in weight would be partly offset by using flanged and welded construction, i.e., all overlaps and rivet heads would be eliminated. The firebox would weigh approximately the same as the original in copper. A further saving would be made at the trailing end by using direct crown stays in place of girders, with the added advantages of making the crown easier to keep clean and having better water circulation.

It was found that, by making the alterations described above in the barrel, that it would be possible to fit 85 x 1.75 in. tubes (an increase of ten) and increase the heating surface to 426 sq. ft. (46 sq. ft. more). Alternatively with 1.625 inch tubes we could have had 480 sq. ft. However, for any given length of tube the effectiveness of the front 30% to 50% diminishes rapidly as the diameter is reduced beyond a certain ratio. At ten feet long, fewer tubes, but of 2 in. diameter, would probably make the most steam.

The Ffestiniog engineers were convinced that the boiler for Lyd should be "superheated" (even at greater first, and increased maintenance, cost), primarily with a view to extending the locomotive's range between water stops. This would inevitably mean reducing the evaporative heating surface, but due to the small barrel diameter this effect would have been excessive if the tube pattern was disrupted by four large flues.

Whilst stuck at this point, we were fortunate to benefit from the advice of Professor John Sharpe. This resulted in the introduction of eight small flues with concentric elements. These cause less disruption to the tube pattern and, if the area of the annulus around each element is in the correct ratio to the area across each small tube, then superheat should be possible with minimal loss of evaporation.

Similar superheaters are used on model locomotives and can be extended through into the firebox as radiant (and powerful) elements. We have this option but with coal burning we might melt them!

The end result has been 57 as against 75 x 1.75 in. tubes, reducing the evaporative heating surface area by 9 sq. ft. but increasing the total to 401 sq. ft. (by 21 sq. ft.). Since superheat counts in theory as 50% greater value, then our 30 sq. ft. becomes 45 sq. ft. making a total equivalent heating surface of 416 sq. ft.

The boiler tube layout has been optimised for oil firing. A prototype oil burner to a special design has been tried out experimentally on one of the FR locomotives and appears most promising. When oil-fired, the locomotive should be one carriage stronger on the WHR (two on the FR). However, by 2009, the firing arrangements were in something of a state of flux as a result of the surge in fuel oil prices. The many, and sometimes political, factors in energy pricing meant that, for example, coal prices could follow oil prices for the longer term.

The atomising system devised for Lyd, based on an air supply rather than solely on steam, should lead to a significant economy in oil consumption of up to 20%. The advantages of air atomising of the fuel as against steam are:

- **higher flame temperature, the major fraction of the steam being generated through radiant heat in the firebox.**
- **greatly enhanced draughting through the tubes, due to reduced volumes of products of combustion, PLUS steam that has been superheated in the flame.**
- **better atomisation due to improved nozzle layout, quicker ignition, hence longer flame path and more perfect combustion.**
- **less water used from the limited tank supply, and less steam drawn from the limited heating surface of the boiler.**

Due to the very much freer gas flow in the tubes (the free gas area being the major limiting factor in this boiler) it is possible to fit a vernier-controlled damper to the ashpan (fitted for either coal or oil burning) so that combustion can be adjusted precisely to suit the particular fuel oil in use, and the working conditions. This damper will also allow the locomotive to be moved on shed with the fire out and without cold air being drawn through the boiler. A further advantage of the air atomising system is that a larger blast pipe orifice is possible (whether Lempor or conventional), which results in lower exhaust back pressure and hence greater tractive power. The mechanical air pump will use less energy than the equivalent steam jet to do the same job. When the locomotive is stationary an automatic change-over valve will allow the atomiser to tick-over on steam.

~ Smokebox Steam and Exhaust Pipes ~

A glance at the steam and exhaust arrangements for Lew showed not only some very tight bends in rather too small diameter pipes, but also some very expensive cast and machined components (See foldout page at end of book). With modern fabrication techniques it has been possible to overcome most of these defects and at the same time produce an exhaust kordina (junction of the two exhaust pipes) that will prevent the steam released from one cylinder causing back pressure on the other.

~ Valve Gear ~

Years of speculation in various books and railway journals has suggested that the valve events on the L&B locomotives were faulty and that this was the reason for their possibly not coming up to expectations in performance. However, as can be seen from the figures produced both by the working model and also by computer simulation (Appendix 2), steam distribution is good and Joy's valve gear is known to give crisp valve events. It has also been believed that the erratic exhaust beats often heard from Joy-fitted locomotives, was due to inadequacies of the design. It would appear that this is not the case, but rather that the effect results from poor maintenance. Once the valves are set, and we have provided an external method for adjusting the valve spindles with ease, the steam distribution will remain satisfactory until the axlebox springs settle, which will significantly alter the situation. It is therefore essential that these are adjusted regularly, especially during the bedding in of new springs. This will not be difficult to do with the setting gauge which will be provided.

It is true that as the locomotive rolls the valve events will be affected, but the overall effect on efficiency should be minimal. From personal observation, it appears that the sound of the exhaust beat is more related to variations in the point of release than the cut-off and, as can be seen, this tends to be more constant than the cut-off which can be uneven especially in back gear.

~ Pony Trucks ~

The original design of the leading and trailing pony trucks allowed the locomotives to roll alarmingly when taking a sharp curve, such as the points at the entry to station loops. The Oakwood video of the L&BR includes a scene which shows one of the Manning, Wardle engines leaving Chelfham Viaduct and lurching heavily as it takes the left hand point into the platform road. Similar to the debate over the performance of the Joy valve gear, has been the question of whether the L&B locomotives rolled excessively. There are several contemporary references to rolling but some of those commenting were not used to the behaviour of narrow gauge locomotives at speed. Most people would find the movements of standard-gauge engines unnerving: rather like sailors being completely at home on lurching boats and rolling ships, whilst land-lubbers soon become unwell!

Not having direct experience of riding the locomotives, I (James Evans) have had to work from the original drawings for Lew and decide which features might require attention and fortunately, since both the height and width of the locomotive are well within reasonable limits for the two-foot gauge, thought has been focused on the trucks which are well out of sight. The first requirement was to alter the swing to allow for curves of less than two chains radius, rather than three or more, so as to allow Lyd to run on the Ffestiniog, the only suitable line at project launch and a major reason for Lyd's subsequent adoption.

This modification sounds simple but, due to the frame stretchers over the trucks extending down between the wheels, these fouled their backs. They therefore had to be completely reworked to bring the load bearing face above the wheel flanges at their highest limit, but without altering the visible top profile. Having done this, careful scrutiny of the suspension showed that the trucks contributed very little to the roll stability of the locomotives and, due to the excessively stiff transverse leaf springs (required because the load on the trailing truck was much more suited to a four wheel bogie), they must have added to any slipping problems wherever there were any track imperfections. In this respect Lyn, with her greater axle load (only two coupled axles) and compensated trucks may well have been more sure-footed than the Manning, Wardles, even though she was lighter (see Table of comparative locomotive dimensions, Appendix 1). Compensation between a truck and its adjacent coupled axle ensures that the load is always distributed between them in the same ratio.

It was therefore decided to adopt the longest possible coil springs, rather like those fitted to the trailing trucks of the Bulleid pacifics, with the object of maintaining a more constant weight distribution between axles, and adding to roll resistance. These have been provided with adjustable rubber bump-stops to control any tendency to pitching.

Side control, which is by springs, has been altered only to allow for the greater swing, though friction pads have been added in the light of Ffestiniog experience with Linda and Blanche, to regulate yawing.

~ *Sanding Gear* ~

The sanding gear of the original engines was fed from two cast tanks mounted on the boiler top. Their location had the advantage of helping to keep the sand warm and dry. A small modification has been made to the sand valves with a view to fully emptying the boxes.

~ *Cab Profile* ~

The structure gauge of the Ffestiniog Railway is limited in its height by the Garnedd tunnel, a little way beyond Tan y Bwlch. The unstable rock formation makes it unlikely that the tunnel could be enlarged without great expenditure. Whilst fitted with a mock-up of the cab profile Lyd was towed through the tunnel and it has been possible to design a modified gutter profile which will be acceptable on the FR and which, by the addition of a bolt-on section each side, can be converted back to the original L&BR appearance when in service on the WHR or visiting the L&BR or other railways.

- V -
Building Lyd

In the late 1980s and early 1990s, the relatively primitive copiers then available made it a traumatic business to copy the fragile set of Eastleigh drawings. Achieving this did however make available a complete set of working drawings. Armed with these, an order for frame plates was placed on 11th May 1992, a lot of metal for modest expense. This was followed by the cylinders, technically very difficult, and needing complete redrawing mainly for pattern-making reasons. The amount of drawing required for the actual production of almost every component proved to be a major challenge (especially for James since he had had no training in engineering drawing!). Next, a full set of tyres was manufactured (with the support of Mike Hart, a director of FR Co.) in South Africa, to Ffestiniog profiles.

An article by Editor Handel Kardas in the April 1995 Railway World featured Lyd and launched the Festiniog Railway Trust's involvement and support. The work would be transferred to Boston Lodge on the Festiniog Railway and would enjoy access to the skills and facilities of the organisation. An appeal to the public to support the project financially was helped by the charitable status of the FR Trust, donations under covenant (now replaced by Gift Aid) being eligible for the reclaim of income tax at the standard rate. The Project Leader would be FR volunteer Paul Lewin (now the General Manager of the F&WHR) and the Design Co-

By 2002 Lyd had become a wheeled 0-6-0 chassis. On 29th May it was in Boston Lodge works yard in the charge of single Fairlie Taliesin. (Jon Whalley)

ordinator would be James Evans, supported by the FR's Chief Engineer, Jon Whalley. Paul persuaded the FR to accept the proposal to organise a Drive an Engine programme on the FR and WHR as part of the fund raising for Lyd. By July 1996 Paul was able to announce that 65% of the estimated funding required had been raised or committed and that the project was on track for completion by 2001. A spare nameplate was raffled in 1995 and won by Mrs Margaret Connell, she and her husband also taking two copper medal units in Lyd funding.

Meanwhile, in July 1995, the frame plates had been moved to Boston Lodge together with the driving wheels, slide bar support brackets, plus the patterns for the pony truck wheels and driving axleboxes. Two whistles had been machined and completed by project treasurer Bill Edmondson and one of them raffled to raise publicity in Steam Railway Magazine. In August 1997 it was reported that a steam roller in Devon which had carried the chimney first mounted on L&BR Yeo, had been secured by James Evans. The chimney proved to be too fragile for regular use on Lyd, however, although it may appear on special occasions! At Boston Lodge the frame plates had

In July 2002 the pony trucks were taking shape. This view shows the nest of four coil springs forming the suspension mounted on the fabricated channel section which will be occupied by the truck axle. (Jon Whalley)

An important test for the pony trucks is the ability to tolerate twist. In this view the movement has been achieved by building up the left hand railhead with steel strip. (Jon Whalley)

Above: *The boiler was completed by Israel Newton in 2002. This front end view of the inner firebox shows the pattern of tubes and flues.* (Paul Lewin)

Right: *Paul Lewin (right) and James Evans celebrate the completion of* LYD'S *new boiler at Israel Newton's workshop during Autumn 2002.* (Valerie Lewin)

Below: *On 2nd November 2002 the completed boiler was delivered to the FR's Minffordd yard where it was carefully lowered into* LYD'S *frames.* (Paul Lewin)

The many casting required to build LYD *needed patterns to be made. Mr Donald Bateman made his skills as a pattern maker available to the project. He is seen here with patterns for the cab window frames.* (David Payling)

February 2004 marked the completion of LYD'S *chimney, crowned by its cast brass cap.* (James Evans)

By 2004 a substantial number of casting patterns had been created. They are stored for future needs at James Evans' workshop in Cornwall. (Paul Lewin)

been drilled ready for erection, and the frame stays machined. Fund raising was supplemented during the May 1997 Bank Holiday by a Drive a Garratt programme at Glan y Pwll where NG/G16 No.138 had been assembled for steam testing prior to delivery to the WHR at Dinas.

In April 1998 it was possible to report that the frame plates had been bolted together in September of the previous year and, as such, Lyd now existed as a locomotive. Meanwhile, in Cornwall the side tanks had been completed at St Austell during March and painted and lettered in temporary Southern Railway livery. At Boston Lodge there was progress towards a wheeled chassis. The forgings for the fly cranks were unusual in that the original design called for the crank pins to be part of the forging. This was to overcome the potential difficulty of securing the pins in the limited depth of metal of the cranks. The fly cranks had been designed in this way by Manning, Wardle to keep the cylinders as close to the frame plates as possible. Unfortunately, one of these forgings proved to have a serious crack in the crank pin area and had to be replaced. Couplings with an authentic L&BR appearance but compatible with FR couplers had been made. Another two major items completed were the patterns and steel castings for the stretchers for the pony trucks. Design work on the axleboxes and the boiler continued.

In August 2000 the wheeled chassis was complete, although without pony trucks at this stage. Brake gear was being fabricated to facilitate the movement of the chassis around the works. The boiler design was complete and made provision for either oil or coal firing. There was to be a steam dryer (providing modest superheat) based on two rows of flues. The tender from Israel Newton of Bradford was

In May 2004 Lyd was displayed at York during the Railfest 200 Exhibition. James Evans puts the finishing touches to the locomotive before visitors arrive. (Martin Cowgill)

The coupling rods were machined to their final tolerances at Boston Lodge in February 2005. (Jon Whalley)

successful and an order placed on 4th December. The chassis had its first run on the FR main line when it was towed as an 0-6-0T to Minffordd for storage. During 2001 the coupling rods and connecting rods were being machined, together with the cylinder covers. By August 2002 the project had spent an estimated £100,000, plus £35,000 for the boiler. At the beginning of November that year the completed boiler was given its hydraulic test at Newton's workshop. It was delivered to Minffordd yard on 7th November where it was craned into Lyd's frames. The locomotive, now complete with its boiler, was then towed down the FR main line to Boston Lodge.

Attention now turned to the pony trucks and Lyd was transferred to Glan y Pwll for temporary storage. The journey was made very cautiously behind Criccieth Castle and no gauging difficulties were discovered on the way to Tan y Bwlch. A short distance beyond this station is Garnedd tunnel, whose portals are the limiting factor in the FR's structure gauge. Lyd's chassis had been fitted with a frame to represent the outline of the cab and this cleared the portal at Garnedd with one inch to spare!

Component manufacture had continued in the workshops both at Boston Lodge and elsewhere. Newly completed items included the crinoline rings to support the boiler cladding. The cladding sheets were also rolled to shape before painting. Ashpan parts and the smokebox door components were also completed. The patterns and cores for the safety valves were made at the Cerdic Foundry in Chard, the same foundry which had cast Lyd's cylinders ten years before. A joint project with the Bluebell Railway's workshop team came to fruition with the casting and machining of a Maunsell displacement lubricator. The crossheads, piston rods and slide bars were all complete and the pistons cast, ready for machining. The tallow cocks attached to the front cylinder covers of the locomotive had been cast. The pony truck side-control springs and housings were being redesigned. A new chimney complete with a cast brass cap was fabricated in James's workshop.

The machining of the balanced slide valves was complete by May 2007. (James Evans)

On New Years Eve 2007 Lyd was towed through Penrhyndeudraeth on its way to Winter storage at Glan y Pwll. Workshop space at The Lodge was required urgently for the rebuilding of NG/G16 Garratt No.87. (Chris Parry)

The machined oil ways and the top oil fillers are visible on the cast dies. The dies slide up and down inside the slide shaft with each revolution of the driving wheels, transferring movement to the slide valve rods. (Jon Whalley)

The curved slide castings can be seen here, together with their connection to the reach rod. The angle of the slide shaft sets the forward or reverse gear. (Jon Whalley)

In 2004 Lyd was invited to appear at the week-long York Railfest 200 and was the subject of much interest. Useful new friendships were made and donations and subscriptions were accepted, bringing forward the completion date for the engine. At about this time agreement was reached with the Lynton & Barnstaple Association to share access to the ever-increasing stock of patterns on a mutually beneficial financial basis. On the workshop front, pressure and vacuum gauges to the original pattern had been completed by Bob Payne and work was in progress with the backhead-mounted injectors. Investigation of the techniques for manufacture of the Joy valve gear slides continued.

By March 2006 it was possible to report completion of the steam dryer, fabricated in stainless steel. On the locomotive the pistons, rings and rods, together with the crossheads (and their newly-machined lubricating oil pots) and slide bars, plus the connecting rods, big ends and coupling rods were all in place. On the drawing board, a design for the steam manifold for control of the oil burner supply was in hand. A design had been produced for the exhaust steam pipes and the seals for their entry to the bottom of the smokebox. Also completed were drawings for the piston and valve glands, the fuel oil tanks and the cab construction details to FR structure gauge clearance. At this point the project to refurbish and restore an ex-SAR NG/G16 Garratt locomotive, No.87, for the opening of the WHR into Porthmadog began to take up all the available space in the Boston Lodge erecting shop, plus much of the time available in the machine shop. Storage in the old FR running shed at Boston Lodge became unavailable in 2008 owing to the restoration and re-roofing of the building and so Lyd was towed to dry storage at Glan y Pwll on 31st December 2007.

Fortunately, during 2008, it was possible to continue to make progress with drawings, workshop tasks and pattern making. Patterns and castings were

The reach rod connects the slide shaft to the reversing lever in the cab. Its very constricted route between the bunker and side tank and the boiler cladding is evident. (Jon Whalley)

made for the locomotive's vacuum brake cylinders and pistons. Balanced slide valves were cast and machined together with four brass oil boxes for the lubrication of the two valve gear curved slides. At last, in January 2009, Welsh Highland Garratt No.87 was out-shopped, freeing space in the erecting shop for work on Lyd to resume. What remained to be done? The locomotive was more complete than it appeared at first sight. Many of the complex castings and forgings had been made or acquired and were in store ready for the final assembly. These included: regulator, safety valves, vacuum ejector, water and pressure gauges, back head clack valves and many other small valves and fittings.

The cladding and insulation of the boiler had been started and needed to be completed so that the boiler and smokebox could be fixed permanently in the frames. Most of the machining of the motion was complete. This included manufacture of the slides for the Joy valve gear –the first to be made for many years. Other machining jobs included the auxiliary steam manifolds, the truck side-control gear and the injectors which were at an advanced state of design. The fabrications for the smokebox door; vacuum brake cylinders and reservoir and the sand boxes were awaiting completion. The ash pan and fire bed carriers needed finishing. Sheet metal work included the cab and running boards and the two fuel tanks. There was also the pipe-work including main steam pipes and exhaust pipes

By December 2009 the final push to complete the locomotive was under way. This was the scene in the erecting shop at Boston Lodge. (David Payling)

A closer look at the locomotive in December 2009 shows some detail of the regulator J-pipe within the dome, the safety valve castings and the cast sand boxes waiting to be fitted. (David Payling)

This smokebox view shows the eight-element steam drier assembly in place. (Jon Whalley)

This August 2009 view shows the spark arresting screens and the blast pipe and blower ring in position. (Jon Whalley)

in the smokebox.

Finally, there were the brass window frames for the cab and the brass beading for the boiler backhead, the latter having been completed by James as a homework project. The polished brass cover for the dome had been completed by Alan Keef at Ross. Lyd would also need the paperwork to be completed for Type Approval. This document includes the details of all engineering modifications to the original design.

Most of the major and many of the minor components were thus now ready for fitting. At this momentous point it was decided to allocate the work of Bob Yates, one of the fitting and welding staff at Boston Lodge, to the completion of erecting the locomotive. Bob was assisted in the work by his father David, a long standing Boston Lodge member

This cab view shows the Maunsell displacement lubricator, diagonally to the right below the window. It is unique among locomotives on the FR, all others having mechanical lubrication. (Roger Dimmick)

Lyd's works plate shows her to be Festiniog Railway No.14, the names of the project and construction team being woven together to read Lew, her L&BR predecessor. (Roger Dimmick)

In June 2009, before the final fabrication of the cab, Lyd made a further gauging trip up the FR main line. The lack of clearance at the mouth of Garnedd Tunnel is shown here. (Jon Whalley)

On 8th July 2010 Bob Yates was about to fit the final components of Lyd*'s valve gear, the right hand return crank rod and its crank.* (Roger Dimmick)

The completed valve gear, 8th July 2010. (Roger Dimmick)

of staff and volunteer. As the months of 2009 went by, progress became rapid. But then, as so often in the past, commercial prerogatives intervened. The workshops won the contract to build a third replica Pickering carriage for the Welshpool and Llanfair Light Railway. In October Bob's time was transferred to the welding of the underframe. Fortunately this was complete by January 2010, allowing work to resume on Lyd and also freeing some much needed erecting shop space.

Bob and David were now able to work full time on making and fitting the countless items which made up the final assembly. They found that the clearances available for components were very tight. Despite the clean, uncluttered external appearance of the MW locomotives there are few square inches to spare. An example proved to be the tortuous route required for the reach rod along the boiler and firebox sides from the reverser in the cab to the weigh shaft. Others included the linkages for the drain cocks, hidden away behind the valve gear. Finding a route for the sand delivery pipes also proved a challenge!

On time, Lyd was out in the Spring sunshine on 1st May and proved to be a star turn at the FR's Gala weekend. The valve gear and many fittings still awaited completion and adjustment but the locomotive was steamed for the first time. In a brief ceremony Lyd was blessed by the Bishop of Bangor. With the gala over, the race was on to complete the many pipe runs for auxiliary steam, air and vacuum, plus the all-important Joy valve gear. Fitting out was completed in July and the valve gear pins sent away for specialist hardening.

Main Photo: *The Bishop of Bangor prepares to bless the locomotive on Sunday May 2nd 2010.* (Roger Dimmick)

Inset: *On Wednesday 5th August 2010 the final component was fitted, Lyd's regulator handle. In keeping with tradition this honour was given to Boston Lodge's newest employee, Dawn Wolfe.* (Roger Dimmick)

- VI -
Commissioning Lyd

On 5th August 2010, with cylinder cocks roaring, Lyd *moved under her own power for the first time.* (Roger Dimmick)

Paul Lewin writes: "………How many of us have looked at photographs of the old Lynton and Barnstaple Manning Wardle 2-6-2T engines and wondered just what it would be like to work on the footplate of one of them? Although there are many tales of the performance of these locomotives, what is missing from the history books is a first-hand account by a driver or fireman of what the originals were like to work on.

Of course, Lyd is a little different from the originals but there is no doubting the family likeness. Something that immediately becomes apparent when working with Lyd is why the crews appear on photographs in certain characteristic poses. When driving and preparing Lyd the crews of today find the same comfortable working positions and vantage points used by the L&BR crews a hundred years ago.

On 5th August 2010, Lyd was rolled out of No.3 Road of the Boston Lodge Erecting Shop and gently warmed up. The burner in the firebox produced a bright clean fire and a hollow sound, when compared with the Laidlaw-Drew burners fitted as standard to FR oil-burning locomotives. Steam was raised slowly with the burner only used for a few minutes at a time at first, to ensure that heat was propagated through the long boiler in an even and progressive way.

The regulator handle, traditionally the last item to be fitted to a new locomotive, was yet to go on. The honour of fitting this final part was bestowed upon the newest employee at Boston Lodge Works - Dawn Wolfe.

As the boiler pressure rose, the routine of oiling up and greasing took place, including the preparation of the hydrostatic lubricator. All FR engines have mechanical lubricators, so some consultation of the manual was required to ensure we got this procedure right! As the lubricator gurgled into life, sounding not dissimilar to a cappuccino machine, steam appeared at the tailpipes of the drain cocks and the cylinders warmed for the first time.

The feeling of a steam locomotive seeming alive is a familiar one to all enginemen, and something often mentioned when people try to describe their experience of steam engines. On this day, Lyd felt alive for the first time, and I must admit I felt a lump in my throat at this point. The overriding feeling was one of great responsibility, and of wanting to ensure that the first few chuffs weren't followed by a loud and expensive bang! Check, check and check again was therefore the order of the moment. I wandered down the yard checking that all points were set correctly for the first move towards Pen Cob and walked back to the locomotive going through a mental checklist for one last time.

I climbed aboard, and everyone seemed to step back a few paces. I found I was reminding myself out loud of everything I needed to do. With the ejector on, full reverse gear and cocks wide open, I asked Bob Yates, leading fitter on the construction of Lyd, and fireman for the day, to wind the handbrake off. I placed my hand on the regulator, which slipped open to the full extent of the small pilot valve, with steam roaring out of the drain cocks. Lyd didn't move. I nudged the regulator towards the main valve but it was very stiff, so I chose instead to ease the drain cocks to partly closed, to give a little extra push, knowing that the cylinders were already very warm. This did the trick and she finally started to move. I instantly pulled the drain cocks wide open again and we were off down the yard.

After just a few chuffs I had to shut off. I went straight to the brake and made a full application. However the brakes seemed very poor on vacuum so Bob wound them on by hand, which did the trick. Clearly something wasn't quite right but we had proved we could stop, so we decided to move again. We ran down the yard a bit further, pretty much rolling all the way on the grade. Now I tried forward gear. Regulator open again, with the drain cocks partially closed and we were off. At last, after fifteen years of hard work, we heard a hearty and rather pleasant chuff emerging from the chimney of an L&BR 2-6-2T!

Delighted I hopped off the footplate and asked Jon Whalley, our Chief Engineer, to run her up and down. Jon had been involved throughout the project and was clearly pleased to see Lyd moving successfully under her own steam. I asked him to drive her up and down the yard a bit so I could stand back and watch her moving. It was a truly emotional moment, comparable with the arrival of a new-born child. I quietly busied myself in the back of the engine shed for a few minutes to avoid the need to talk to anyone but all the time looking back across the yard at a truly magnificent sight.

We spent the next hour or two moving round the yard and checking everything. Full back gear produced a knock felt through the frames and the brakes appeared to need some bedding in, but otherwise things seemed fine. Appropriately for the FR, one of the first jobs Lyd was called upon to do was to move a couple of slate wagons around the yard.

Having been double-checked by experienced fitters, as a supplement to our own inspection, Lyd was deemed to be fit to run over to Porthmadog to pick up some fuel. The single-line token was taken out and we were soon on the Cob. We ran with a very small regulator opening with ears straining to hear every last sound but there was nothing to hear. We closed the drain cocks fully and then a familiar sound was heard. It brought back memories of driving the other FR 2-6-2T, Mountaineer. There was a bump as the back truck passed over a rail joint, then a pause followed by three bumps in rapid succession as the drivers went over and then another pause before the front truck did the same. A blast on the GWR whistle announced our presence at the fixed distant and we tried the vacuum brake several times to start the bedding-in process.

My first impression was one of stability, sturdiness and power. The ride was very smooth - almost as steady as a double Fairlie, with no evidence of the rolling mentioned in old accounts of the L&B locomotives. The valve gear produced a clean square beat with a crisp cut off, although it would be some days before we heard the sound of the engine under any significant load. The regulator has a very small pilot valve, which is handy for shunting, and the main valve is very stiff indeed - particularly with 200psi behind it (the originals had 160psi boilers). The L&B whistle is very shrill indeed and we later fitted it with a steam choke to reduce the volume a little.

We arrived at Porthmadog, triumphant! Plenty of people had worked out what was going on and shutters clicked all around. We were soon out checking every bearing to make sure there was nothing untoward. The cross heads were warm, but the slide bars showed a good light grey oil film so we carried on. We took a little fuel but needed to move out of the way of an inbound train. So it was off into the loop to stand trying to look relaxed as a trainload of passengers arrived at the station.

Although we had only a couple of miles 'on the clock' we felt confident enough to offer Control help with the shunt. They needed five cars taking off the set and putting in the sidings. We hooked onto passenger stock for the first time. Much to my amazement, Lyd moved these five cars almost as if there was nothing there. We used only the small pilot, or easing, valve. However, we decided against pushing our luck and took the engine back to the workshops. It had already been a long day with a lot of adrenalin flowing. Her Majesty's Railway Inspector was due the next day. Lyd was to be the last locomotive in the UK to be approved under the ROTS regulations (Railways and Other Transport Systems) and we needed to be in good shape for his visit. As we considered a trip to the pub to celebrate, someone mentioned how cool it was that Boston Lodge Works had built new steam engines in the 19th, 20th and, now, the 21st centuries. Who else can make such a claim?

Lyd's second day started early as we prepared for the visit of the Inspector. With no 'grandfather rights' Lyd was effectively a new design of locomotive, differing from the L&BR originals in several respects that are critical to safe operation. We had originally applied for approval under the ROTS regulations over ten years previously. This was to allow inspections to take place regularly during construction, as a means of avoiding a fatal flaw emerging at the last moment. It was nonetheless a day tinged with apprehension.

We were out on the line early, to get some miles of running-in done to help bed-in the brakes, so as to be able to complete a successful brake test. We headed out up the line to Gwyndy Bank, well away from any crossings, so we could get up a little more speed and make vigorous brake applications. Everything seemed fine, apart from the brakes! We clocked up a number of minor points for attention but nothing more than would be expected with any new machine - such as finding the lamp didn't fit on the rear bracket. An injector water valve handle needed turning through ninety degrees, and so on. We got a message that the Inspector was half an hour away and ran back down to Boston Lodge to do a final check over and grab a much-needed cup of coffee.

The first part of the inspection was spent reviewing the drawings. There are a lot of them, and many have been updated to clarify dimensions. In the case of the redesigned pony trucks they had been completely redrawn. Various calculations were questioned and the files proved that they were in order. The boiler design paperwork was also discussed in detail, as the boiler is another item that differs significantly from the original.

With these points dealt with, we moved out to the locomotive and went over it top to bottom and end to end. I ran through the driver's preparation list with the Inspector. Bob talked through the fireman's preparation list as well. It was a useful way to go over all aspects of the machine in a methodical fashion. The first problem spotted was that access to the top of the tanks was not easy for crew members with short legs. Generally speaking, there is no need to go up top during the day, but preparation and disposal do require it. The second snag, noted soon afterwards, was the lack of a dipstick in the fuel tanks. For historical reasons this is a standard fitting on FR locomotives. However, with fuel meters now fitted to all tank wagons we had felt it unnecessary for Lyd. We reconsidered this decision and provisionally added it to the snag list.

With no other obvious issues we prepared to run Lyd on the main line. We started with a light engine run. The Inspector was invited to try various

On Sunday 8th August Merddin Emrys speeds past Rhiw Goch with the 10.15 departure from Porthmadog as LYD *rests in the loop with her first loaded test train.* (Roger Dimmick)

controls to establish that they functioned correctly and were appropriately located. Questions were asked such as: 'Whilst sanding in difficult driving conditions can the driver still maintain a forward view whilst operating the sander control?' - and so on. The first few miles were uneventful although the knock on the driver's side remained. We could see an 'obvious' cause, and a perceived binding-spot on the motion was eased - but found to make no difference! A driving style avoiding full reverse gear avoided the worst of it and bought us some thinking time.

The brakes appeared to be improving, but were borderline in brake tests, so we knew something was amiss. Curiously the handbrake, which worked the same mechanism as the vacuum cylinders, was going from strength to strength.

As we were sitting in Minffordd Station waiting for the Up service train to clear the section, a golden moment presented itself. Whilst standing back and admiring the engine a group of intending passengers came by. The usual questions ensued - 'Where does the train go?' and so on. This was quickly followed by: 'How old is this engine, then?' We retaliated with 'How old do you think?' This produced answers varying from seventy to a hundred years. How delightful to be able to say 'Just one day old, actually!'

It was agreed that a further brake test was needed, with the locomotive hauling vacuum-fitted stock. This was duly arranged, with our Chief Mechanical Engineer at the regulator and answering further questions as needed. At the conclusion of the test, just five points stood between us and approval. These were: provision of access to the tanks tops, dipsticks for the tanks, improvement in the dynamic brake performance, a loaded test to establish the rated load and the ability of the locomotive to start this load on the normal operating gradient, and finally a static brake test to prove that the engine brake could hold a train on this gradient. We felt quite satisfied with such a short list, and set about making plans to sort it out. The first improvement came very quickly as our quiet and thoughtful Carriage Maintenance Fitter asked permission to take the vacuum cylinders apart, as he had a feeling he knew what was causing the problem. It turned out he was right. He discovered that the rolling ring in the vacuum cylinder was not sealing properly against the cylinder wall and piston, allowing the vacuum to leak off. He effected a temporary repair to prove the point and the next test run found the brakes in excellent order. The repair was made permanent during the next on-shed day.

Steps to provide access to the tank tops would have been easy to fit. However, the problem was that the original engines did not have them and they were bound to be so obvious as to make Lyd look very different. The solution arrived when Bob Yates showed us the temporary step he had used during construction. It hung on the web between the water tank and the smokebox. We liked the idea of a removable step and a pair was made. The crew can take them off if they wish, but many photographs of the locomotive over the past few months show them left in place all day. This will probably change once the engine is painted in its final livery, when they will become more obvious against a green background

Load testing, and the ironing out of problems, required mileage more than anything else. However, it was August and the FR was busy with peak summer traffic. Finding a path that did not interrupt the service, and finding a set of carriages that was free to be used as a test train, required some very early starts. Some runs were truncated by late running, as we stopped to check bearings and re-apply oil. From time to time, we ran to section time and the locomotive gave a glimpse of what it might be capable of doing once fully run in. On the Sunday after the HMRI Inspection, we ran a filming special to get some early publicity shots. The train

Lyd's naming took place on 11th September 2010. From the left the participants were: James Hamlin, Rebecca and Jon Whalley, Paul Lewin, Mary Martin, Val and daughter Millie Lewin, James Evans, Bob and David Yates. (Roger Dimmick)

crew included: Driver, Paul Lewin; Fireman, James Hamlin; Travelling Fitter, Bob Yates and Guard, David Payling. Whilst stabled in a siding at Rhiw Goch we were unable to relight the fire. We had run out of fuel! Despite our efforts to check everything in great detail we had missed the obvious, and at this point decided that dipsticks were definitely a good idea. We also decided that putting the engine in the shed and spending the afternoon in a deckchair was also not a bad idea. As we cleaned up Lyd prior to doing just that, we couldn't help but notice that the crossheads were not running any cooler. Also the heavy black fouling of the used bearing oil was getting worse. Something was amiss.

Some time previously we had been "running in" another engine that was visiting the FR, fresh from overhaul. After hours and hours of bedding in, we could not get the axle boxes to run cool. We oiled, checked and mollycoddled the engine to no avail. With little time left we decided to try a kill-or-cure approach and ran the engine at a good speed with a heavy load. In short, this brutal treatment worked and, after a couple of miles and a further round of oiling, she started to run cool. Similarly we were not sure what was going on with Lyd's crossheads. They were measured accurately and found to be in good order. They were fine tuned just in case and the operation of the lubrication system verified. We therefore decided to push her harder in the hope of detecting the problem more readily.

At that point we had run Lyd with six cars for load testing, and it was becoming apparent that she was capable of much more. The FR's Double Fairlies are rated at twelve cars and the Penrhyn Ladies, Linda and Blanche, at eight. We hoped that Lyd would be a competent eight-car engine as this would make it of maximum use in the FR fleet. We therefore waited until the end of the day's service and picked up the stock from a terminating service train. We counted nine cars and thought that should do. We ran at line speed and accelerated at the normal firm rate associated with the FR services. Steaming was borderline and timekeeping was good. However, the crossheads were not at all happy. The used oil contained large quantities of black and copper coloured flecks. We cleaned them off at station stops and re-lubricated but soon they were back. This was not a problem that any amount of bedding in or oiling would fix. We turned back at Tanygrisiau, some eleven miles out, happy with the

After the naming ceremony LYD spent the rest of the day shuttling the vintage FR carriage set between Beddgelert and Pont Croesor. (Roger Dimmick)

knowledge that we had an engine that would pull nine cars and still had something to spare but we made a decision that the crossheads were coming off for re-working before the engine was needed in traffic for the various launches around the country.

Lyd's crossheads are made of cast iron and run on steel slide bars. Nothing unusual there - but the grades of the materials used, combined with the loading being experienced by the components was causing the problem. A metallurgist friend explained that the harder elements of the cast iron were digging into the steel surface and an oil film was not being established. The material dislodged was being rolled into small balls which melted and welded themselves to the slide bars. This then hardened and cut back into the cast iron cross head. In turn, this exposed a new surface in the iron and the whole process started again. In one hundred and twenty five miles we had been losing one thousandth of an inch of metal per mile. That is the kind of wear one might expect over 30-50,000 miles of normal hard running. Fortunately, we had seen this effect before on Merddin Emrys. We knew not to try quick fixes, but rather to rework the crossheads with white metal fused onto the bearing surfaces and to re-machine them to the original tolerances.

As is often the case, we were up against a deadline with just a week to sort it out and get though the running-in phase again - needing mileage and time. On the positive side, we had concluded the tests required by HMRI. A letter authorising the use of the locomotive arrived the day before the first planned passenger train. Team work is something which means playing your best people at the right time. Lyd needed experienced fitters with stamina and machinists who could quickly assimilate a lot of information and set up machines right first time. Boston Lodge has them and they were at their best in that last week of August. They knew it was important to a lot of people to get Lyd right. They said they could do it in a week if left to get on with it and they delivered. I must admit to being incredibly proud of the Boston Lodge team as they worked to achieve the required result. It had been a tough call to take the valve gear down for this work, but it had paid off. Lyd now runs with a fine grey sheen on the slide bars indicative of a perfect oil film between the two parts.

We made it to the official launch on the Welsh Highland Railway with a little room left for manoeuvre, but the engine ran incredibly well. Even that persistent knocking on the driver's side was identified by one of the team watching the locomotive's every move. He spotted the point where the motion bracket needed the slightest amount of easing.

Commissioning the locomotive is just the end of the first part of an ongoing adventure. It is a rare privilege to be deeply involved in an adventure only made possible by the determination, persistence and passion of a committed team............"

Lyd climbs Gwyndy bank with a test train on 16th August 2010. (Roger Dimmick)

- VII -
Early Running

The WHR Superpower Weekend at Dinas had been selected as the day for Lyd's inaugural run. The passengers on this special train were Lyd project supporters who had patiently waited so long for this day. With Lyd at its head, the vintage FR carriage set left Dinas at 8.55 a.m. and headed South towards Beddgelert. At Castell Cidwm a brief stop had to be made to allow a fence post and wire to be removed from behind Lyd's pony truck –deposited on the track earlier by straying cattle. Lyd managed the WHR grades with ease and arrived on time at Beddgelert. After being detached from the train the locomotive stood in the Down platform road where, at 11.00a.m, it was named by Val Lewin and Mary Martin, wives of the project leaders. For the rest of the day Lyd hauled the vintage carriage set between Beddgelert and Pont Croesor, to the delight of the many passengers.

After its launch at the Superpower weekend, Lyd travelled to its native West Country, first stop being the Launceston Steam Railway's Gala weekend on 18-19 September. Here Lyd hauled a train of Launceston carriages over the 2½ miles to Newmills. Air braking capability was achieved by the attachment of quarry Hunslet 0-4-0ST Lilian to the rear of the train. The following weekend, 24th to 26th September, Lyd was to be in action at the L&BR Association's Woody Bay Autumn Gala. Paul Lewin kept a diary:

"...........At Launceston Lyd was the biggest engine the Steam Railway had ever accommodated. The photos show it to look a little out of scale for the infrastructure. It was only to be expected that the engine would create a lot of interest on arrival in Cornwall. That was indeed the case, and James Evans certainly experienced the pressure of being in the media spotlight for a couple of days.

Passenger services with Lyd were scheduled for the Sunday, and people arrived in considerable numbers. Just as at Beddgelert, it proved to be a gathering of old friends. Operations at Launceston were limited as it was not possible for Lyd to negotiate the run round loop. It had been designed with quarry Hunslets in mind. As a result, topping and tailing of the carriages was the only workable solution. The sun shone and the atmosphere was like a summer garden party at an English country house. Tea was drunk, cakes eaten, admiration of the locomotive's form was expressed, and the minutiae of the engineering were earnestly discussed. James was driving his engine on his home railway and a happier man I have never seen.

The visit to Launceston was brief. A day was allowed for Lyd to cool down, and the following Tuesday she was drained for the move to Barnstaple. She was to appear in Barnstaple High Street for publicity purposes before broadly tracking the route of the L&B on her way to Woody Bay. On Friday morning at Woody Bay the engine stood simmering outside the shed in early morning light, surrounded by gentle wisps of steam. Seeing this made the significance of the occasion sink in. In just a few minutes we would be hearing the sound of an L&B whistle on the L&B for the first time in seventy five years. Ronald Redman, president of the Narrow Gauge Railway Society, and a member since its

LYD pauses at Launceston station between trains during the Gala weekend of 18-19 September. (David Payling)

inception sixty years previously, had said there were two things he thought he would never see. One was trains in the Aberglaslyn pass. The second was a working Lynton and Barnstaple engine. His words were echoing in my ears!

It was also a strange feeling to see the original L&B coach No15 (now FR buffet car No14) back in Devon. Left behind at Snapper Halt as a garden shed when the L&B was scrapped, it had been taken to North Wales in pieces in 1959 and had become one of the FR's most loved carriages. It had been in the last L&B train in 1935, hauled by Yeo and Lew. Now, it was back once more, for a short while at least. Friday was devoted to route learning, testing both locomotive and train, and working with the press. We had brought both ex-L&BR car No.15 and FR Observation saloon No.102 from the FR, as the L&B needed the capacity. Also, Lyd just doesn't look right with smaller stock. The transport costs had been tripled by moving the carriages too, but it was worth it and hopefully next time some original profile stock will be available from the L&B's own rebuilding program. Using FR stock also overcame the issue of vacuum brakes, with which Lyd is fitted, as was the original L&B stock. James Evans, fittingly, had the first opportunity to drive an L&B engine back into Woody Bay and we had already agreed that I would drive the first passenger train the following day.

The atmosphere around the station the next

On Sunday 19th September LYD *approaches Newmills with an afternoon train during the Launceston Steam Railway Gala weekend. The train consists of LSR carriage stock and with air braking supplied by 0-4-0ST Hunslet Lilian at the rear of the train.*
(David Payling)

morning is not easy to put into words. The sun shone on north Devon whilst the rest of the country was hit by squalls, and the whole feel of the place was very sunny indeed. So many people – all so very happy. The park and ride arrangements worked well – a good job too, as the station would have been full of cars in the first half hour otherwise. Woody Bay was filled with people feeling that it was good to be alive – you could sense it from quite a distance as the feeling really did seem to embrace you. There were many well-known faces, and many people who simply came and said thank you for making this happen. I think we developed a defence mechanism of being very jovial and light about the whole thing. Otherwise we would have spent all day with lumps in our throats, thinking about the enormity of it all.

Lyd behaved impeccably, as you would imagine. She never really got hot in the whole time she was there, although she did make a good attempt at straightening out some of the tight curves in the run-round loops. We smiled to have found that an L&B engine did not fit the new L&B loading gauge! Several notes were made about correcting that, and a particularly troublesome point lever disappeared. The hard working PW lads sorted out a height problem in Killington Lane platform, for which we were most grateful. No doubt it will all have been fully sorted by our next visit. Provision of fuel was worked out by a couple of L&B volunteers who were very clever with pumps and tanks. We were impressed, as

James Evans and Paul Lewin pause before attaching the commemorative wreath to Lyd's smokebox door. Like the original wreath on the bufferstop at Barnstaple Town in September 1935 it carried an inscription. This one read: "Perchance it is not dead but sleepeth... Perchance it's awake!" (Tony Nicholson)

Below: *Recreating a scene on the pre-war L&BR Lyd and its train of FR Observation car No.102 and ex-L&BR Carriage No.15 (now rebuilt as FR buffet car No.14) approach Woody Bay from Killington Lane.* (Tony Nicholson)

During the L&BR Association's Gala Weekend of 25th-26th September 2010, Lyd and train rest between duties at Woody Bay station. (David Payling)

we were with the copious amounts of tea and snacks that came our way. We were also blessed with the help of a number of Lyd team supporters, especially including Julie Stirland, who came along each day to clean the locomotive. Their efforts ensured that Lyd was spotless all the time.

As the weekend progressed, we managed to get an incredible number of people either onto the footplate or to ride in the train with us. We'd always wanted to share the engine with like-minded folk and it was great to do so. John Prideaux, Chairman of the FR Company, and author of a book on the L&B many years ago, was persuaded back to Woody Bay to open the gala weekend, and rode on the footplate with a big smile on his face. Lucky heather from the Aberglaslyn Pass was presented to the crew by Mary Martin and carried on the headlamp, something often seen in old photos of the line. A wreath of bronze chrysanthemums was also carried, similar to the one laid on the buffer stops at Barnstaple when

A view from the cab window as LYD *and train approach Woody Bay on the L&BR on Sunday 26th September, 2010.* (David Payling)

Driver Paul Lewin leans from the cab as Lyd *rounds Dduallt Farm Curve and approaches Bunny's Crossing with its first passenger duty on the Ffestiniog Railway – the 10.15 Porthmadog to Blaenau Ffestiniog on Friday 6th October 2010.* (Roger Dimmick)

At Harbour Station on 30th October 2010, it's all smiles as ex-SAR Garratt No.87 and station pilot locomotive Lyd *stand side by side after the arrival of the Bronze Supporters train from Caernarfon.* (Roger Dimmick)

In the shade filled Aberglaslyn Pass, LYD and TALIESIN roll down the 1 in 60 gradient from Tunnel 2 with the nine-car "Platelayers' Special" to Rhyd Ddu on 31st October 2010. (Roger Dimmick)

the line closed in 1935. It had carried a card reading 'Perchance it is not dead but sleepeth', and there was certainly no sleeping when Lyd was working up the grade......."

Lyd was back in North Wales for the FR's Vintage weekend on 8-10 October. It was in charge of the 10.15 a.m. Friday departure from Porthmadog to Blaenau. This was the first public train on the FR to be hauled by the locomotive. It then shared the haulage of trains, double heading with other FR engines, for the rest of the weekend. Finally 30th October saw the first trains from Caernarfon over the rebuilt WHR to reach Harbour Station, running through the streets of Porthmadog and over the Britannia Bridge. The two trains that day were specials for the Gold, Silver and Bronze subscribers to the WHR Project. Lyd acted as the station pilot locomotive whose role was to attach to the rear of the train as it came to a halt on the embankment (Cob) and then to haul the train back into the platform. To assist the departure of the trains, the same manoeuvre was repeated in reverse, Lyd hauling the long train out of the platform onto the embankment. Finally, to acknowledge the huge contribution of the WHR track gangs, a special train was arranged for them and their families on 31st October. It ran from Harbour Station over the Britannia Bridge and over the WHR to Rhyd Ddu and return. The train was headed by Taliesin and Lyd and once again was a great success.

Later in the year Lyd was the subject of a photo charter on the WHR. In December, with the engine in plain black livery, and no prospect of an early space in the paint shop, some Lyd supporters requested that the engine be temporarily turned out in BR mixed traffic livery. Appropriate grey and red lining was therefore applied together with 'cycling lion' transfers. However, the project has always aimed to finish the engine in 1925 Southern livery and this will be applied in the near future.

Winter time gives an opportunity for workshop attention. It is hoped to be able now to make progress with the many details yet to be completed on Lyd.

- VIII -
CONCLUSIONS

The three original Manning Wardles were sold at auction for scrap in 1935 for £34-50 (approximately £2650 in today's money) and cut up in approximately one week! The timescale (18 years) and cost (£370,000, plus countless volunteer hours) of recreating one of these engines for the 21st century makes for a sobering comparison.

In terms of evaluation it is too early to reach final conclusions about Lyd's capabilities. However some factors have already emerged. Close study of the Lew drawings has revealed a number of errors, some quite serious, others amusing and a few which may possibly be useful for historical research. In the latter category it was found that OPEN and SHUT are shown reversed on the regulator quadrant. This may be a simple draughtsman's error or could perhaps suggest that the design was conceived with left hand drive in mind and the regulator link was then altered, but not the quadrant wording?

Again, the original drawings (and the works photographs and the engraving in Engineering Magazine, 1897) show the vacuum ejector exhaust pipe running through the lower part of both sand boxes. This would not have been possible if they had been made to the drawings! This is because, as originally drawn, the duct through the front and rear sand boxes was too small for the diameter of the exhaust pipe. Also, there are dimensions in the cylinders that do not add up and there is an obvious mistake in the diameter of the main steam pipe which was made too small.

Returning to the engine's capabilities, the slightly modified cab roof allows the locomotive to run comfortably over the FR and WHR. Only the outer sections of the cowcatchers have had to be omitted at the moment as they are too close to platform edges. These sections are also omitted in a works photo of Exe. The weight of the engine is 20.75 tons empty and 25 tons in working order, clarifying errors in a number of publications down the years which stated Lew to be some 4¼ tons heavier than it was! The leading truck carries 3.5 tons, the leading coupled axle 6.5 tons, driving axle 6.0 tons and trailing coupled axle 5.0 tons, the trailing pony truck carrying 4.0 tons (the latter two being subject to further adjustment). Lyd is thus 1¾ tons heavier than Lew. The valve gear has proved to give a very even beat, suggesting good settings and no problems have so far been discovered.

Although the stability of the ride of the original Manning Wardles has been criticised, Lyd has shown no sign of either pitching or rolling and sharp curvature is accommodated smoothly. This is possibly the result of the redesigned truck suspension. For steaming the draughting and smokebox arrangements have worked satisfactorily. In the firebox the newly designed oil burner has also proved effective, although refinement is now in progress, aiming especially to achieve good combustion at small steaming rates. It is also intended to use coal firing at some future point.

It is too early to give firm figures for passenger car loadings on the two railways. The L&BR ruling gradient is more severe at 1 in 50 than the FR's 1 in 80, but less so than then WHR's formidable 1 in 40. Assignment of regular passenger loads will take into account the ability to keep time without undue effort or excessive wear. Test trains on the FR have already reached a maximum of 9 cars but an everyday load of 8 cars may be more reasonable and perhaps some 5 cars over the WHR. Such loads suit operations on the FR where a six-car set is standard, supplemented by additional cars as required. On the WHR, however, the longer train sets of some 8 cars means that Lyd is likely to appear on lighter duties or double heading with another locomotive. Lyd's capability is in either case, however, far greater than the 4 cars (plus any freight cars picked up at wayside stations) allowed for the original engines over the L&BR. What we certainly can say is that the locomotive has already proved itself to be a promising and valuable member of the F&WHR motive power fleet, with its eager power and free running.

On Sunday 8th August 2010 Lyd worked a load test train of six carriages to Cei Mawr, seen here at Lead Mine Curve. (Roger Dimmick)

- IX -
Acknowledgements

The Project Team of Paul Lewin, James Evans and Jon Whalley offer their grateful thanks to the many subscribers to the Lyd fund. Without their support over many years this undertaking would have been impossible to bring to a successful conclusion. We are grateful too for the continued support and commitment of the FR Co. board and the FR Trust for providing a charitable trust framework for the management of the project.

There have been very many contributors to the practical side of the project, far too numerous to be identified by name. However, some who must be mentioned include: Bill Edmondson, treasurer to the project since its inception (and maker of Lyd's whistle and tallow cups!). Gordon Bartlett has kept subscribers in touch with progress through his role as Editor of the Lyddite Newsletter. Thanks also go to the many skilled volunteers and paid staff of the workshops at Boston Lodge who have striven to ensure that the quality of Lyd will be a match for its L&BR predecessors. We owe a special debt of gratitude to the late David Perrin for his inspiring painting of Lyd with its train at Minffordd station. This not only helped to raise public awareness of the project, but raised our spirits at a time when we had a long way to go!

On matters relating to the boiler, thanks are due to Professor John Sharpe (superheaters), Dave Holt and Graham White (early stages of boiler design), Nigel Day (combustion) and Roger Pridham (boiler construction details). Bob Yates (together with his father, David) carried out the final erecting of Lyd, resolving the many problems of fitting so much into so little space. On design and in the workshop we thank: Mike Balderstone (modifications to ease manufacture of the curved links of the valve gear), Simon Bowditch (valve gear simulations), Donald Bateman (Pattern making), Humphrey Davies (advice on patterns and castings), John Cashmore (injector design and advice), Geoff Cox (helped to drill the frame plates), John Hill (machinist and networker), Paul Singleton (metallurgy of motion parts), Bob Maxwell (machining of cross-heads, Bob Payne and Ernie Binley-Smith (precision engineers and machinists), David Lewin (help with many tricky items), Roger James (machining the slide shaft oil boxes), Llyr ap Iolo (rebuild of Dreadnought ejector), the NRM (detail of name and works plates), Craftmaster Paints, Les Driver (motion clearance problems at Launceston) and the Taliesin Team at Boston Lodge (construction of many components). Nigel Bowman (Launceston Steam Railway) has been a willing consultant on all things mechanical and on the boiler.

Steve Philips has researched details not shown in the SR drawings and David Bailey has carried out livery research. We thank Colin Pealling (published researches) and Geoff Horsman (expert knowledge of Hunslet Engine Co).

Last, but not least, we thank Roger Dimmick (our in-house photographer) and the many organisations and photographers who have made their work available for this book. Acknowledgement of their work appears with the illustrations. Lastly, we apologise sincerely to anyone who has helped the Project and whom we have accidentally overlooked.

- X -
Appendices

1. *Leading Dimensions Of L&Br and Selected Locomotives*

	Manning Wardle *Lew*	FR *Lyd*	Baldwin *Lyn*	Hunslet** *Blanche*	Hunslet *Russell*
Date Built	1925	2009	1898	1893	1906
Number Built	1	1	1	2	1
Gauge (in.)	23½	23½	23½	23½	23¼
Wheel Arrangement	2-6-2T	2-6-2T	2-4-2T	2-4-0TT	2-6-2T
Builders Number	2042	14	15965	589	901
Cylinders	10 ½ x 16	10 ½ x 16	10 x 16	10 ½ x 12	10¾ x 15
Driving Wheels (in.)	33	33	33	26	28
Leading Wheels (in.)	24	24	22	19	18
Trailing Wheels (in.)	24	24	22	-	18
Heating Surface (Tubes) sq. ft.	343	334	339	222 (+23 flues)	345
Heating Surface (Firebox) sq. ft.	37	37	40	28.75	36
Heating Surface Total sq. ft.	380	401 Incl. s/h	379.2	295	381
Grate Area (sq. ft.)	8.85	8.74	7.7	5.2	6 ¼
Ratio GA:HS	1:43	1:47	1:49	1:56	1:61
Tube Length . & diam., in	120 x 1.75	120 x 1.75 (& 3.5in.)	105¾ x 1.75	106 x 1.75 (& 5 in.)	101 x 1.625
Free Gas Area. (sq. In)	144.5	143	147	130	144

	Manning Wardle *Lew*	FR *Lyd*	Baldwin *Lyn*	Hunslet** *Blanche*	Hunslet *Russell*
Ratio GA:FGA	8.82	8.8	7.5	5.76	6.25
Boiler Pressure psi	160	200	180	160	160
Water Capacity (Gallons)	550	550	664	270 (saddle tank)	440
Coal Capacity	35 (cu.ft.)	35 (cu. ft)	6¼ (cwt)	Tender oil tank	14½ (cwt)
Length over Buffer Beams (ft.)	27	27	23ft. 6in.	26ft. 2in.	19ft. 5½in.
Coupled Wheelbase	6 ft. 6 in.	6ft 6in	5ft	5ft	5 ft. 6 in.
Total Wheelbase	17 ft. 9 in.	17ft 9in	17ft 7in	8ft. 9in.	15ft. 6in.
Width over platform	6 ft. 7 in.	6ft 7in	7ft 2in	5ft. 7in.	6ft. 5in.
Maximum Height	8ft. 11in.	8ft 11in	8ft 11in	7ft. 11 7/8in.	8ft. 11in.
Weight (working order) tons	*	25	22	15.3	20
Adhesive Weight lb.	39,312	39,200+	28,500	29120	?
Tractive Effort (lb.) (85% wp.)	7269	9087	7418	6920	7425
Ratio AW:TE	5.4:1	4.3:1	3.84:1	4.21:1	?
Max. Axle Load lb.	13,832	14,560+	14,250	17920	?

* A weight of 27¼ tons in working order has often been quoted for these engines. However, a weight of 22 tons in working order was quoted by Mr FW Chanter (Engineer to the L&BR) in the book "Narrow Gauge Railways – Two Feet and Under," by LS Robertson (1898), reprinted by Plateway Press, 1988. In the Southern Railway General Arrangement drawing the figure for Lew is given as 23.34 tons (see Conclusions chapter for Lyd's weights).

** The figures for Blanche are as running currently on the FR. Dimensions for Linda differ from Blanche in certain details such as the tender, the slide valve cylinder castings, etc.

\+ Current figures, but subject to future adjustment.

Abbreviations: Sq. - *square* GA - *Grate Area* HS - *Heating Surface Area* Diam. - *Diameter* FGA - *Free gas area* AW - *Adhesive Weight* TE - *Tractive Effort*

2. A Computer Modelling Study of the Manning, Wardle - SR Valve Gear Drawings for Lew and a Full-Size Mechanical Model Made from Those Dimensions

The reversing gear is as built with the reach rod at 120.25 in. centres, the same as the distance between the weigh shaft and reversing lever (NOT circa. 120.5 in. which would be correct for a 4° slope). When the reverser is vertical, the weigh shaft arm is circa -1° back from vertically upwards.

The curved slide is set at approximately 3° when the arm is at 0° (i.e. vertical), giving its centre as 2 ¾ in. below the weigh shaft (as shown on SR Drawing No. 28). The mechanical model (see photograph p.22) shows that circa 2° gives mid-gear, therefore with the lever vertical and the weigh-shaft arm at -1°, the curved slide at 2°, the centre of radius is approximately 2 in. below the weigh shaft. This geometry gives the best relationship between the lever and the weigh shaft. Full gear at the reverser pin joint is +/- 4.125 in., approximately equal to +/- 11.5° degrees.

FROM THE FULL-SIZE MODEL

Maximum cut-offs:			
Full Fore Gear:	+14.75° at weigh shaft (from mid gear of -1°)		
FP (Front Port)	81% cut off	94% release	5% compression
BP (Back Port)	83% cut-off	95% release	6% compression
Average	82% cut-off (N.B – this was beyond the accurate limit of the model)		
Full Back Gear:	-15° at the weigh shaft (from mid gear at -1°)		
FP	84% cut-off	95% release	7% compression
BP	79.5% cut-off	93% release	5% compression
Average	81.75% cut-off		
Minimum Cut-Offs:			
Fore Gear:	+4.125° (as above)		
FP	34% cut-off	74% release	27% *
BP	32% cut-off	74% release	25% *
Average	33% cut-off		
Back Gear:			
FP	33% cut-off	73% release	28% *
BP	29% cut-off	71% release	27% *
Average	31% cut-off		

At 29-34% minimum cut-offs the figures for compression seem high, although the ports are quite long and wide. However, with the increased working pressure of 200 psi it might be better to limit the minimum cut-off to 35% and allow the engine to be worked on slightly closed regulator. This should give less steam chest pressure on the valves and slightly hotter steam due to the expansion through the regulator.

With experience of running and testing under load the reverser rack could be altered at some future date, if required. Alterntively the lands between notches 2, 3 and 4 could be reduced by one sixteenth of an inch. This would give the following results:

Minimum Cut-Offs:

Fore Gear:	+4.25°		
FP	36% cut-off	75% release	25% compression
BP	34% cut-off	74% release	25% compression
Average:	35% cut-off		
Back Gear	-4.5°		
FP	36% cut-off	75% release	26% compression
BP	32% cut-off	74% release	25% compression
Average:	34% cut-off		

N.B The locomotive could possibly sound off beat in back gear.

This view from the servicing pit highlights the detail of LYD'S Joy valve gear. (Roger Dimmick)

3.Boiler Construction Details

Boiler Shell and Superheater

All inner and outer firebox seams were to be flanged and butt welded, including the fire hole, saving weight, avoiding dry-lap seams and giving maximum flexibility. In this and some other respects it was decided to depart somewhat from BS 2790 which is not specifically a locomotive boiler standard. Due to the known lack of weight towards the leading end of the locomotive, and defects in the performance of the original boilers, it was decided to increase the barrel diameter marginally (3/8 in), within the limits of clearance from other components and within the original cladding profile. Together with no reduction of the diameter of the front section, this allowed the elimination of the blocking ring between the barrel and the smokebox.

Because of the restricted diameter of the smokebox the dryer header and elements were made as a single welded unit and the tubing of stainless steel, the whole item being easily withdrawn through the smokebox door. The front tube plate was riveted into the barrel to reduce the stress-raiser of a square cornered weld, especially with due regard to the unavoidable closeness of the tubes to the edge of the plate. This also allowed for reasonable access to the boiler shell interior for repairs without firebox removal, the latter being welded at the foundation ring. Entry by the dome would be very tight and would require the removal of longitudinal stays.

Stays

On the advice of the boiler makers, the brackets for the short longitudinal stays and the back head doubler plate have been riveted.

Firebox stays have been kept to traditional threaded and headed type except in the case of the crown stays, where special welded ends were required in the outer row each side where they pass through the semi-circular outer wrapper at an acute angle. These enlarged ends are full penetration welded and provided with tell-tale holes to allow easy drilling and tapping at future replacement. Remaining crown stays are threaded and seal welded outside, and threaded, caulked and nutted on the fire side and provided with tell-tale holes.

Palm stays to support the firebox tube plate have not been used (see Manual of Locomotive Engineering, WF Pettigrew and AF Ravenshear, 1899, p.212, and also Steam Locomotive Design Data and Formulae, EA Phillipson, 1936, p.105), since a steel tube plate is used. Research has shown that there is considerable stress caused by the differential expansion between the tube bundle and palm stays, leading to leakage of the adjacent tubes and frequent breakage of the stays. Whilst allowing the tube plate in this area to provide some flexibility to overcome these problems the unsupported area has been minimised by lowering the tube bundle whilst maintaining a reasonable water space and keeping the uppermost throat plate stays as high as possible. The latter have also been specified in monel (nutted on the fireside) whilst eight tubes have been designated as stay tubes of greater wall thickness and beaded over in the firebox (as also are the flue tubes). Monel has been selected for the top row of side stays due to their providing support on the diameter of the semi circular upper casing. No tubes are to be welded due to experience with tube plate cracking under FR oil burning conditions. If absolutely necessary tubes can be beaded and welded, i.e. to the ligament and not round the edge of the holes.

The monel stays are not intended to outlast the

plain steel stays and should be replaced at the same frequency. They are intended to provide superior durability in the areas of highest stress.

Firebox

A brick arch is provided which is to be combined with an air-atomising oil burner, refractory fire bed of reduced air-inlet area and full ash pan with vernier-controlled damper. This system, which has been developed with various trials over a long period, is designed to produce fully controllable combustion and reduced thermal shock within the firebox. All components have been designed for easy removal via the firebox door, if conversion to coal firing is required. However, the tube and steam dryer specification has been calculated with oil burning in mind and it is anticipated that with coal fuel the fire bed will need to be kept thin and bright to produce good results. This will require high standards of technique and, with heavy loads, will lead to spark throwing. The fitting of effective spark arrestors to a boiler with unavoidably poor free gas area, will inevitably result in reduced haulage power of up to about 20%, even with good coal and skilled management.

Other Details

The expansion brackets supporting the boiler on the outer firebox sides, being subjected to extreme loadings due to the torsional resistance of the boiler to the flexing of the long locomotive frames, are riveted so as to avoid long fillet welds which would raise stresses. With the possibility of leakage from the rivets, unseen when the boiler is in the frames, the heads are welded inside the water space.

All remaining details have been subjected to calculation and increased in strength by 25% in line with the working pressure.

All studs in areas subjected to extra wastage are specified in monel. Internal steam pipes for the injectors have been routed from above the firebox to get the hottest steam and shortest pipes. The feed pipes have been routed to the front of the barrel and provided with small holes to discharge the feed above the water surface to guarantee mixing and release of oxygen. These sections are removable for cleaning. The original Gresham & Craven backhead-mounted combination injectors being unsuitable and prohibitively expensive to replicate, new fittings have been designed which look similar. They incorporate clack valves developed at the Launceston Steam railway on Hunslet Velinheli and modern spherical steam valves which do not leak and are easily serviced. The injectors are of standard Ffestiniog Railway Gresham type and mounted where they will not be seen externally but will be in a fully flooded position and visible to the engine crew without leaning outside the cab. We plan also to make our own sets of cones specifically designed for 200p.s.i working pressure and to best current practice.

A deflector plate has been provided above the firebox tube plate to aid circulation over the firebox crown and reduce the mixing of water droplets with steam passing to the dome, always a serious problem with small diameter domed boilers with high working water levels.

5. A Daydream

There was that 1896 exhibition at Newlay with Calthrop, Leeds Forge and Kitson showing off their latest ideas for the narrow gauge railways of the Empire. Manning, Wardle might have been a little worried about the competition and would certainly have noted the novel ideas. They may have had an order, or perhaps only a tentative enquiry, from a similar part of the world? If work was a bit slow, they might well have jumped the gun and started work on what, for them, was a very non-standard design.

Then you have the L&BR directors, concerned at the rising costs of building their railway, agreeing a favourable but perhaps unrealistic contract with Brush in Loughborough. When this falls through, and with their opening date advancing, they make hasty enquiries of all of the other builders. The fact that Manning, Wardle could produce three locomotives to a non-standard design, both quickly and at a good price, makes one think that they had something suitable already on their hands (or in the pipeline!).

It is interesting that only three locomotives were purchased, when it can be seen that, if an order was placed from scratch, at least four would have been required. Also it appears from several details that the design was modified. The weight on the trailing truck suggests that a 2-6-4T arrangement may have been intended. Details in the regulator suggest that left hand drive was planned (as on the Leek & Manifold, for example). The curious cab with its large sliding windows hints at work in a hot climate. The extended cab sides allowed the windows to open and, whilst making a steam trap on cold, damp Exmoor, would not have created a problem in the hot climate of India! Again, the motion guards would have been very useful in a hot and dusty climate.....

.... Well, perhaps?

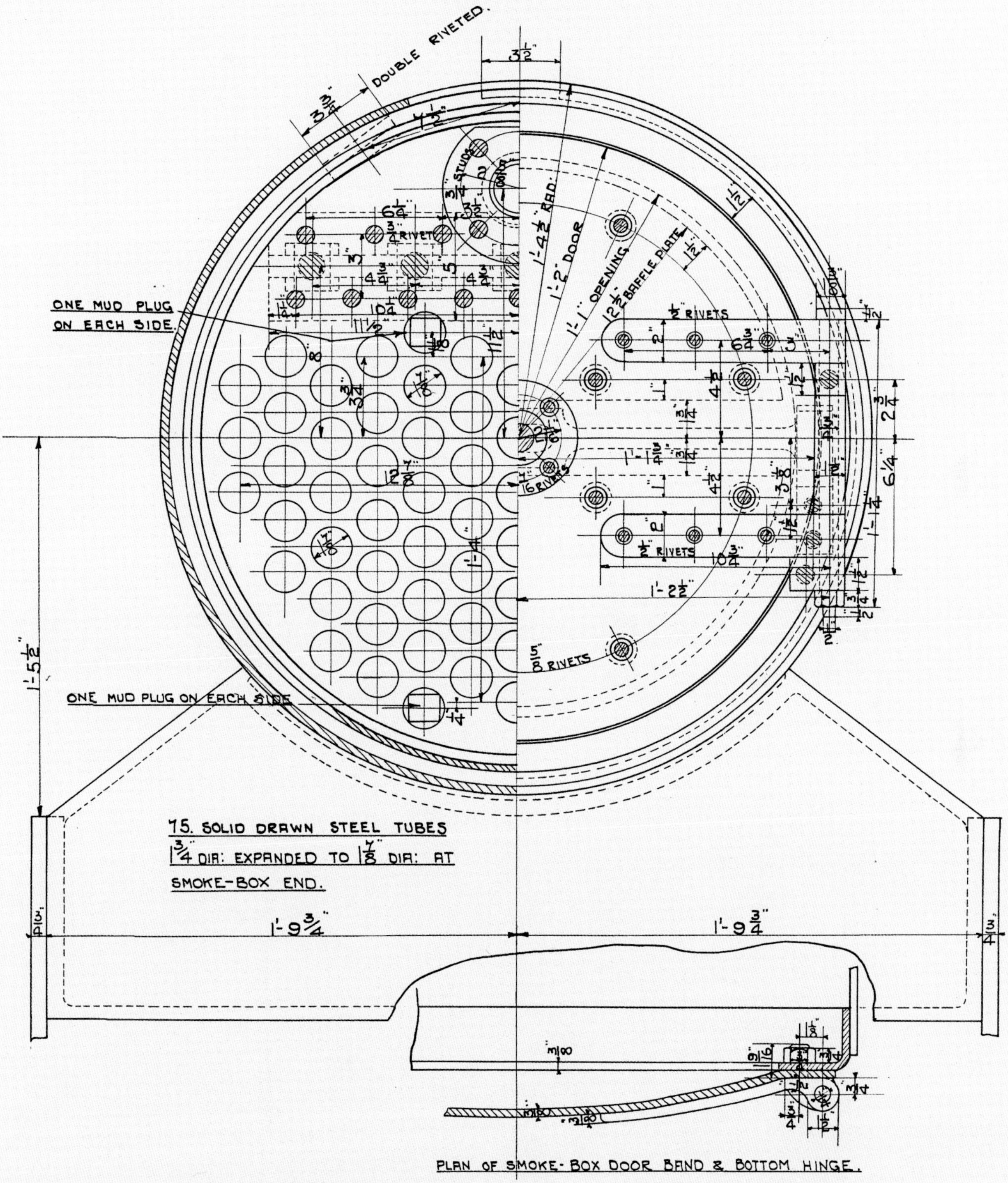
DOUBLE RIVETED.
ONE MUD PLUG
ON EACH SIDE.
ONE MUD PLUG ON EACH SIDE
1'-4½" RAD.
1'-2" DOOR
1'-1" OPENING
12½" BAFFLE PLATE
½" RIVETS
⅝ RIVETS
75. SOLID DRAWN STEEL TUBES
1¾" DIA: EXPANDED TO 1⅞" DIA: AT
SMOKE-BOX END.
1'-9¾"
1'-9¾"
PLAN OF SMOKE-BOX DOOR BAND & BOTTOM HINGE.

SMOKEBOX

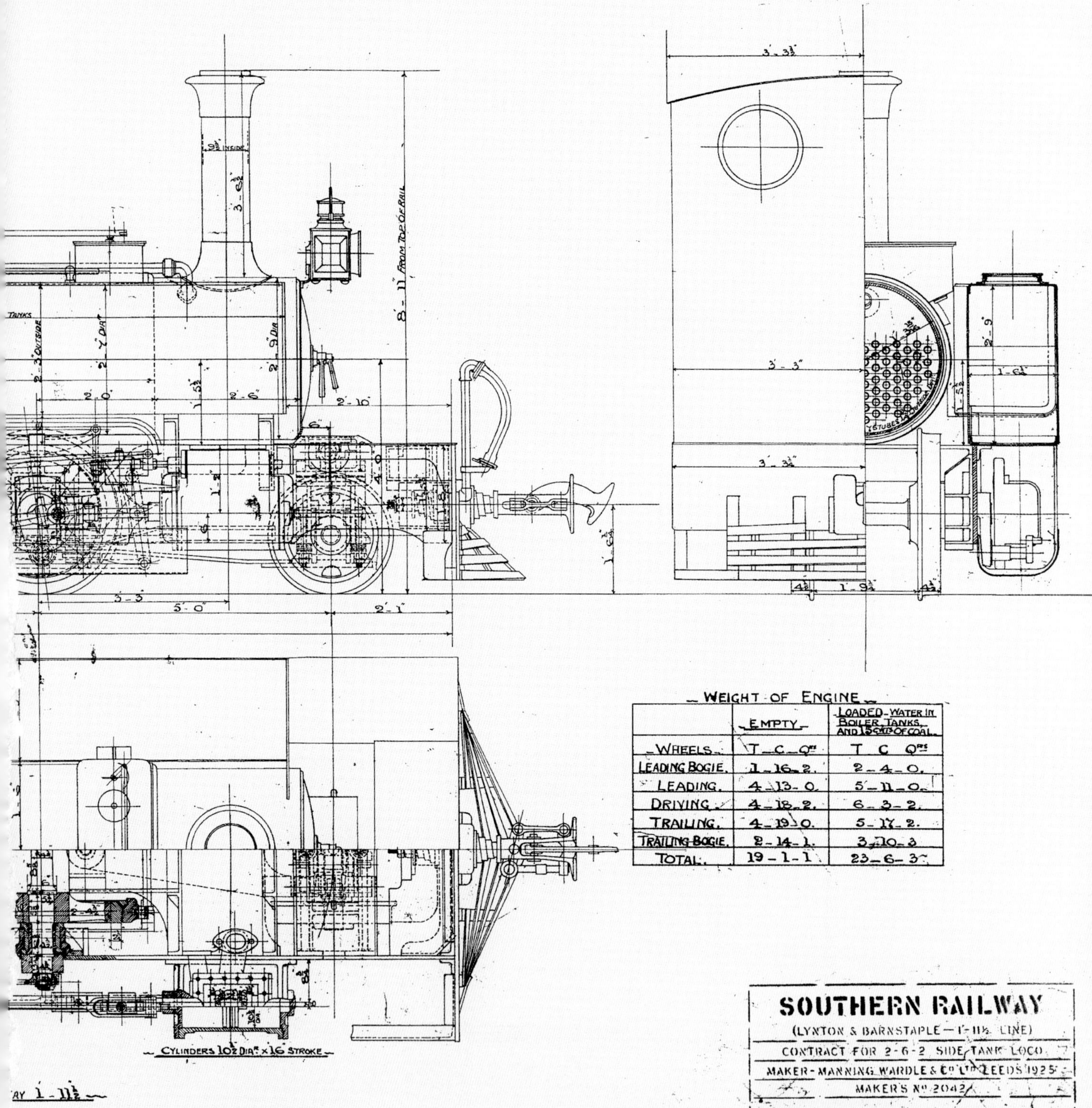

.ENT (Drawing No.2) OF L&BR LOCOMOTIVE

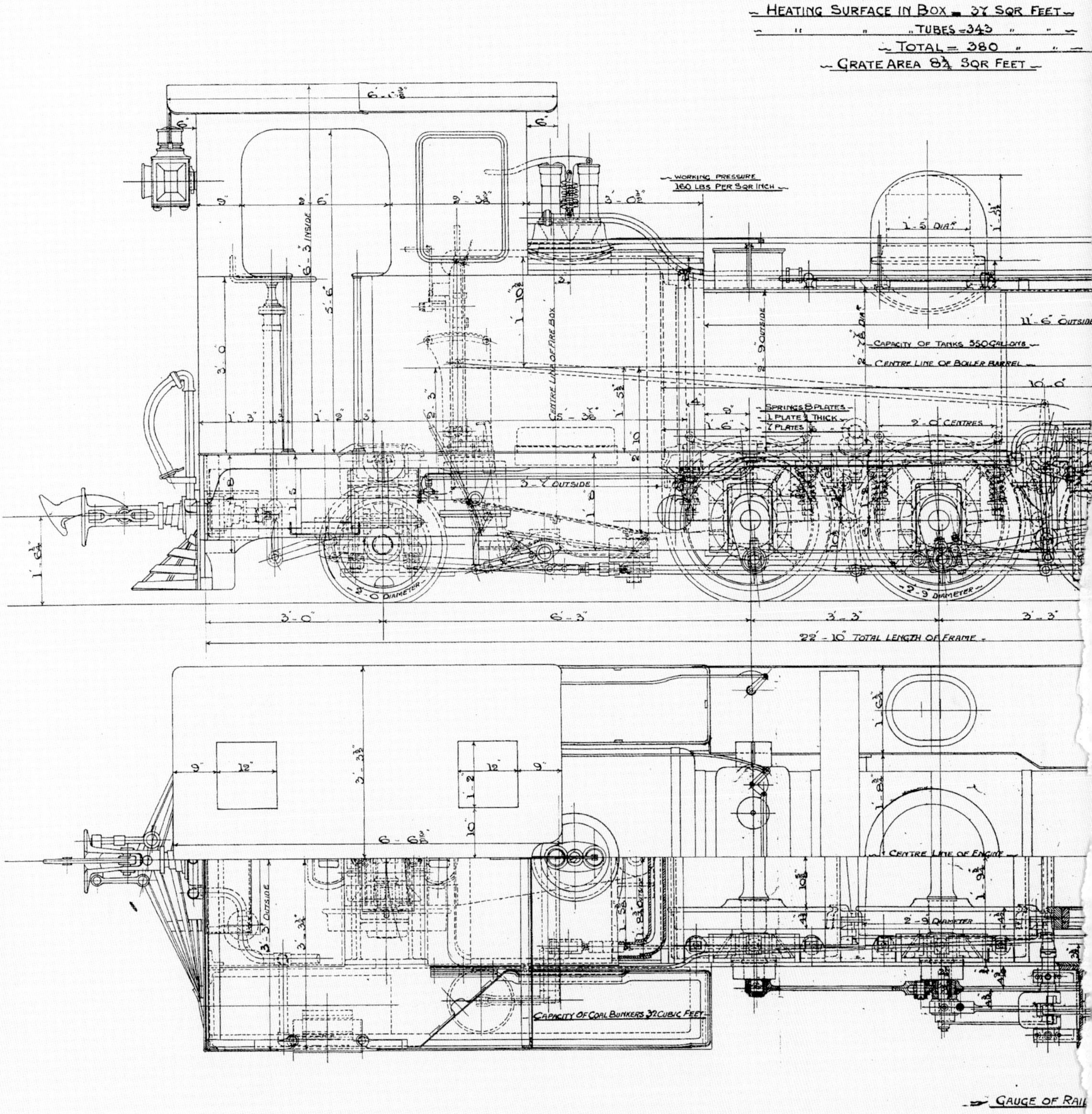

SOUTHERN RAILWAY GENERAL ARRANG

BOSTON LODGE WORKS
LEWIN, YATES
EVANS &
WHALLEY
No: 14 2010